Structural Engineering Formulas

Ilya Mikhelson, Ph.D.
Tyler G. Hicks, P.E.
Illustrations by Lia Mikhelson, M.S.

Second Edition

New York   Chicago   San Francisco
Lisbon   London   Madrid   Mexico City
Milan   New Delhi   San Juan
Seoul   Singapore   Sydney   Toronto

Cataloging-in-Publication Data is on file with the Library of Congress.

McGraw-Hill Education books are available at special quantity discounts to use as premiums and sales promotions, or for use in corporate training programs. To contact a representative please visit the Contact Us page at www.mhprofessional.com.

Structural Engineering Formulas, Second Edition

Copyright © 2013 by McGraw-Hill Education. All rights reserved. Printed in the United States of America. Except as permitted under the United States Copyright Act of 1976, no part of this publication may be reproduced or distributed in any form or by any means, or stored in a data base or retrieval system, without the prior written permission of the publisher.

1 2 3 4 5 6 7 8 9 0   DOC/DOC   1 2 0 9 8 7 6 5 4 3

ISBN 978-0-07-179428-2
MHID 0-07-179428-X

The pages within this book were printed on acid-free paper.

Sponsoring Editor
Larry S. Hager

Acquisitions Coordinator
Bridget Thoreson

Editorial Supervisor
David E. Fogarty

Project Manager
Vastavikta Sharma,
Cenveo® Publisher Services

Copy Editor
Patti Scott

Proofreader
Megha Saini,
Cenveo Publisher Services

Production Supervisor
Pamela A. Pelton

Composition
Cenveo Publisher Services

Art Director, Cover
Jeff Weeks

Information contained in this work has been obtained by McGraw-Hill Education from sources believed to be reliable. However, neither McGraw-Hill Education nor its authors guarantee the accuracy or completeness of any information published herein, and neither McGraw-Hill Education nor its authors shall be responsible for any errors, omissions, or damages arising out of use of this information. This work is published with the understanding that McGraw-Hill Education and its authors are supplying information but are not attempting to render engineering or other professional services. If such services are required, the assistance of an appropriate professional should be sought.

To my wife and son
I. M.

About the Authors
Ilya Mikhelson, Ph.D., had more than 30 years of experience in design, research, and teaching design of bridges, tunnels, subway stations, and buildings. He wrote numerous publications, including *Precast Concrete for Underground Construction, Tunnels, and Subways* and *Building Structures.*

Tyler G. Hicks, P.E., is a consulting engineer and a successful engineering book author. He has worked in plant design and operation in a variety of industries, taught at several engineering schools, and lectured both in the United States and abroad. Mr. Hicks holds a bachelor's degree in Mechanical Engineering from Cooper Union School of Engineering in New York. He is the author of more than 20 books in engineering and related fields, including *Civil Engineering Formulas, Handbook of Mechanical Engineering Calculations,* and *Handbook of Energy Engineering Calculations.*

Contents

Preface to Second Edition xi
Preface to First Edition xiii
Acknowledgments xv
Introduction xvii

Part I Basis of Structural Analysis

1 Stress and Strain: Methods of Analysis 3
    Table 1.1 Stress and Strain: Tension and Compression 5
    Table 1.2 Stress and Strain: Bending 7
    Table 1.3 Stress and Strain: Bending 9
    Table 1.4 Stress and Strain: Combination of Compression (Tension) and Bending 11
    Table 1.5 Stress and Strain: Torsion 13
    Table 1.6 Stress and Strain: Curved Beams 15
    Table 1.7 Stress and Strain: Continuous Deep Beams 17
    Table 1.8 Stress and Strain: Dynamics, Transverse Oscillations of the Beams 19
    Table 1.9 Stress and Strain: Dynamics, Transverse Oscillations of the Beams 21
    Table 1.10 Stress and Strain: Dynamics, Transverse Oscillations of the Beams 23
    Table 1.11 Stress and Strain: Dynamics, Impact 25
    Table 1.12 Stress and Strain: Dynamics, Impact 27

2 Properties of Geometric Sections 29
    Table 2.1 Properties of Geometric Sections: Tension, Compression, and Bending Structures 31
    Table 2.2 Properties of Geometric Sections: Tension, Compression, and Bending Structures 33

Table 2.3	Properties of Geometric Sections: Tension, Compression, and Bending Structures	35
Table 2.4	Properties of Geometric Sections: Tension, Compression, and Bending Structures	37
Table 2.5	Properties of Geometric Sections: Tension, Compression, and Bending Structures	39
Table 2.6	Properties of Geometric Sections: Torsion Structures	41

Part II Statics

3 Beams: Diagrams and Formulas for Various Loading Conditions ... 47

Table 3.1	Simple Beams	49
Table 3.2	Simple Beams	51
Table 3.3	Simple Beams	53
Table 3.4	Simple Beams and Beams Overhanging One Support	55
Table 3.5	Cantilever Beams	57
Table 3.6	Beams Fixed at One End, Supported at Other	59
Table 3.7	Beams Fixed at One End, Supported at Other End	61
Table 3.8	Beams Fixed at Both Ends	63
Table 3.9	Beams Fixed at Both Ends	65
Table 3.10	Continuous Beams	67
Table 3.11	Continuous Beams: Settlement of Support	69
Table 3.12	Simple Beams: Moving Concentrated Loads (General Rules)	71
Table 3.13	Beams: Influence Lines (Examples)	73
Table 3.14	Beams: Influence Lines (Examples)	75
Table 3.15	Beams: Computation of Bending Moment and Shear Using Influence Lines (Examples)	77
Table 3.16	Beams: Computation of Bending Moment and Shear Using Influence Lines (Examples)	79

4 Frames: Diagrams and Formulas for Various Static Loading Conditions 81

- Table 4.1 Frames: Diagrams and Formulas for Various Static Loading Conditions 83
- Table 4.2 Frames: Diagrams and Formulas for Various Static Loading Conditions 85
- Table 4.3 Frames: Diagrams and Formulas for Various Static Loading Conditions 87
- Table 4.4 Frames: Diagrams and Formulas for Various Static Loading Conditions 89
- Table 4.5 Frames: Diagrams and Formulas for Various Static Loading Conditions 91

5 Arches: Diagrams and Formulas for Various Loading Conditions 93

- Table 5.1 Three-Hinged Arches: Support Reactions, Bending Moment, and Axial Force 95
- Table 5.2 Symmetrical Three-Hinged Arches of Any Shape: Formulas for Various Static Loading Conditions 97
- Table 5.3 Symmetrical Three-Hinged Arches of Any Shape: Formulas for Various Static Loading Conditions 99
- Table 5.4 Two-Hinged Parabolic Arches: Formulas for Various Static Loading Conditions 101
- Table 5.5 Two-Hinged Parabolic Arches: Formulas for Various Static Loading Conditions 103
- Table 5.6 Fixed Parabolic Arches: Formulas for Various Static Loading Conditions ... 105
- Table 5.7 Fixed Parabolic Arches: Formulas for Various Static Loading Conditions ... 107
- Table 5.8 Three-Hinged Arches: Influence Lines 109
- Table 5.9 Fixed Parabolic Arches: Influence Lines 111
- Table 5.10 Steel Rope 113

Contents

6 Trusses: Method of Joints and Method of Section Analysis 117
    Table 6.1 Trusses: Method of Joints and Method of Section Analysis 119
    Table 6.2 Trusses: Method of Joints and Method of Section Analysis 121
    Table 6.3 Trusses: Influence Lines (Examples) 123
    Table 6.4 Trusses: Influence Lines (Examples) 125

7 Plates: Bending Moments for Various Support and Loading Conditions 127
    Table 7.1 Rectangular Plates: Bending Moments 129
    Table 7.2 Rectangular Plates: Bending Moments (Uniformly Distributed Load) 131
    Table 7.3 Rectangular Plates: Bending Moments (Uniformly Distributed Load) 133
    Table 7.4 Rectangular Plates: Bending Moments (Uniformly Distributed Load) 135
    Table 7.5 Rectangular Plates Bending Moments (Uniformly Distributed Load) 137
    Table 7.6 Rectangular Plates: Bending Moments and Deflections (Uniformly Distributed Load) 139
    Table 7.7 Rectangular Plates: Bending Moments (Uniformly Varying Load) 141
    Table 7.8 Rectangular Plates: Bending Moments (Uniformly Varying Load) 143
    Table 7.9 Circular Plates: Bending Moments, Shear and Deflection (Uniformly Distributed Load) 145

Part III Soils and Foundations

8 Soils 151
    Table 8.1 Soils: Engineering Properties 153
    Table 8.2 Soils: Weight/Mass and Volume Relationships 155
    Table 8.3 Soils: Stress Distribution 157
    Table 8.4 Soils: Settlement 159
    Table 8.5 Soils: Settlement 161
    Table 8.6 Soils 163
    Table 8.7 Bearing Capacity Analysis 165

Contents

9 Foundations 167
    Table 9.1 Foundations:
                     Direct Foundations 169
    Table 9.2 Foundations 171
    Table 9.3 Foundations 173
    Table 9.4 Foundations 175
    Table 9.5 Foundations: Rigid Continuous
                     Beam Elastically Supported 177
    Table 9.6 Foundations: Rigid Continuous
                     Beam Elastically Supported 179
    Table 9.7 Foundations: Rigid Continuous
                     Beam Elastically Supported 181

Part IV Retaining Structures, Pipes, and Tunnels

10 Retaining Structures 185
    Table 10.1 Retaining Structures: Lateral
                     Earth Pressure on Retaining Walls 187
    Table 10.2 Retaining Structures: Lateral
                     Earth Pressure on Retaining Walls 189
    Table 10.3 Retaining Structures: Lateral
                     Earth Pressure on Retaining Walls 191
    Table 10.4 Retaining Structures: Lateral
                     Earth Pressure on Retaining Walls 193
    Table 10.5 Retaining Structures: Lateral
                     Earth Pressure on Retaining Walls 195
    Table 10.6 Retaining Structures: Lateral Earth
                     Pressure on Braced Sheetings 197
    Table 10.7 Retaining Structures: Cantilever
                     Retaining Walls 199
    Table 10.8 Retaining Structures:
                     Cantilever Sheet Pilings 201
    Table 10.9 Retaining Structures:
                     Anchored Sheet Pile Walls 203

**11 Pipes and Tunnels: Bending Moments
for Various Static Loading Conditions** 205
    Table 11.1 Pipes and Tunnels: Rectangular
                     Cross Section 207
    Table 11.2 Pipes and Tunnels:
                     Rectangular Cross Section 209
    Table 11.3 Pipes and Tunnels:
                     Rectangular Cross Section 211
    Table 11.4 Pipes and Tunnels:
                     Rectangular Cross Section 213

Table 11.5	Pipes and Tunnels: Rectangular Cross Section	215
Table 11.6	Pipes and Tunnels: Circular Cross Section	217
Table 11.7	Pipes and Tunnels: Circular Cross Section	219
Table 11.8	Pipes and Tunnels: Circular Cross Section	221
A	**Quick-Use Conversion Tables**	**225**
B	**Mathematical Formulas: Algebra**	**235**
C	**Mathematical Formulas: Geometry, Solid Bodies**	**239**
D	**Mathematical Formulas: Trigonometry**	**243**
E	**Symbols**	**247**
	Index	**249**

Preface to Second Edition

When Larry Hager, Senior Editor, McGraw-Hill Professional, asked me to revise Ilya Mikhelson's *Structural Engineering Formulas* book, I was awed by the request. I had used this excellent book in my own engineering practice, and I knew the book to be a superb, and highly useful, treatise.

Thinking about the revision, I decided that the best way to update the book was to use the blank left-hand "Notes" pages for new, related content. Doing this would update the book without unduly increasing the page count or the price of the book. Further, I would, where possible, leave room for Notes. This would preserve, to some extent, Dr. Mikhelson's unique idea of leaving room for important comments by the reader. Given engineer's proclivity for making notes about their work, it made sense to leave as much room as possible for note making.

The new content nearly doubles the technical coverage of the book. Some 300+ new formulas have been added, along with 40+ new illustrations. Specific new topics in the Second Edition include the following: strain energy principles; strain energy in structural members; stress-strain relations; stress and strain failure analysis; analysis and design of flat and curved springs; properties of geometric sections of columns; torsion of shafts of various cross sections; shaft twist and torque formulas; beam loading formulas; position of flexural center for different sections; torsion in solid and hollow shafts; safe loads for beams of various types; torsion in structural members; eccentric loading of beams; combined axial and bending loads; computation of fixed-end moments in prismatic beams; continuous beam analyses; curved beam analyses; influence lines; natural circular frequencies and natural periods of vibration of prismatic beams; columns and frames; short columns; elastic flexural buckling of columns; formulas for circular rings and arches; eccentrically curved beams; curved beam position stress factors; reactions of a three-hinged arch; length of cable carrying known loads;

truss design and application; column baseplates; local buckling of plates; bearing plate design; determination of flange plate thickness; determination of stresses in plates; formulas for flat plates; relationship of weight and volume in soils; lateral pressures in soils; forces on retaining walls; lateral pressure of cohesionless soils; allowable soil bearing pressures; allowable loads on piles; toe capacity load on piles; determination of foundation settlement from soil test borings; estimation of structure settlement; Housel's method for determining the foundation footing size; cantilever retaining walls; geosynthetics in retaining wall construction; concrete gravity retaining walls; six types of retaining walls and their soil-pressure variation; stability of a retaining wall; pressure on submerged curved surfaces; flexible and rigid ditch conduit loads; pipe stresses for various load conditions; forces due to pipe bends; and pipe on supports at intervals. Appendix A includes a much-expanded group of conversion factors for USCS and SI unit conversions that will be helpful to every engineer and designer using this book.

In closing, I thank Larry Hager for his excellent guidance during the revision of this book. During preparation of this revision I consulted a number of engineering handbooks that I use in my professional engineering work. These sources are cited throughout this book. It is my sincere hope that if Dr. Mikhelson were to see this revision, he would approve of the added content.

TYLER G. HICKS, P.E.

Preface to First Edition

This reference book is intended for those engaged in an occupation as important as it is interesting—design and analysis of engineering structures. Engineering problems are diverse, and so are the analyses they require. Some are performed with sophisticated computer programs; others call for only a thoughtful application of ready-to-use formulas. In any situation, the information in this compilation should be helpful. It will also aid engineering and architectural students and those studying for licensing examinations.

ILYA MIKHELSON, PH.D.

Acknowledgments

Deep appreciation to Mikhail Bromblin for his unwavering help in preparing the book's illustrations for publication.

The author would also like to express his gratitude to colleagues Nick Ayoub, Tom Sweeney, and Davidas Neghandi for sharing their extensive engineering experience.

Special thanks are given to Larry Hager for his valuable editorial advice.

I. M.

Acknowledgments for Second Edition

The following books have been especially helpful in providing much essential data, and many illustrations, that appear throughout the Second Edition of this book.

> Brockenbrough, Roger L., and Frederick S. Merritt, *Structural Steel Designer's Handbook*, 3d ed., McGraw-Hill, New York, 1999.
>
> Hicks, Tyler G., *Civil Engineering Formulas*, 2d ed., McGraw-Hill Professional, New York, 2009.
>
> Merritt, Frederick S., M. Kent Loftin, and Tonathan T. Ricketts, *Standard Handbook for Civil Engineers*, 4th ed., McGraw-Hill, New York, 1996.
>
> Roark, R. J., *Formulas for Stress and Strain*, 4th ed., McGraw-Hill, Book Company, New York, 1965.

T. G. H.

Introduction

Analysis of structures, regardless of its purpose or complexity, is generally performed in the following order:

- Loads, both permanent (dead loads) and temporary (live loads), acting upon the structure are computed.
- Forces (axis forces, bending moments, shears, torsion moments, etc.) resulting in the structure are determined.
- Stresses in the cross sections of structure elements are found.
- Depending on the analysis method used, the obtained results are compared with allowable or ultimate forces and stresses allowed by norms.

The norms of structural design do not remain constant, but change with the evolving methods of analysis and increasing strength of materials. Furthermore, the norms for design of various structures, such as bridges and buildings, are different. Therefore, the analysis methods provided in this book are limited to the determination of forces and stresses. Likewise, the included properties of materials and soils are approximations and may differ from those accepted in the norms.

All the formulas provided in the book for analysis of structures are based on the elastic theory.

PART I

Basis of Structural Analysis

| **CHAPTER 1** | **CHAPTER 2** |
| Stress and Strain: Methods of Analysis | Properties of Geometric Sections |

CHAPTER 1
Stress and Strain: Methods of Analysis

4 Basis of Structural Analysis

NOTES

Tables 1.1 through 1.12 provide formulas for the determination of stresses in structural elements for various loading conditions. To evaluate the results, it is necessary to compare the computed stresses with existing norm requirements.

Stress and Strain: Methods of Analysis   **5**

TABLE 1.1   Stress and Strain: Tension and Compression

Weight 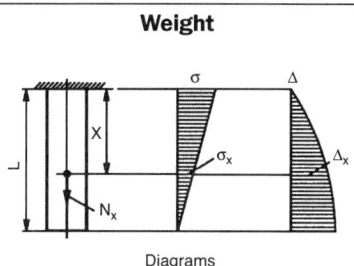 Diagrams	Axial force: $N_x = \gamma A(L-x)$, γ = unit volume weight, A = cross-sectional area. Stresses: $\sigma_x = \dfrac{N_x}{A} = \gamma(L-x)$, $\sigma_{x=0} = \gamma L$, $\sigma_{x=L} = 0$. Deformation: $\Delta_x = \dfrac{\gamma x}{2E}(2L-x)$, $\Delta_{x=0} = 0$, $\Delta_{x=L} = \dfrac{\gamma L^2}{2E} = \dfrac{W^2 L}{2EA}$ $W = \gamma A L$ = weight of beam, E = modulus of elasticity.
Axial force: tension, compression 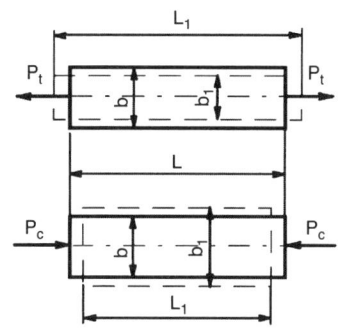	Stresses: $\sigma_t = \dfrac{P_t}{A}$, $\sigma_c = \dfrac{P_c}{A}$. Deformation: $\Delta_L = L - L_1$ (along), $\Delta_b = b - b_1$ (cross), $\varepsilon_L = \dfrac{\pm \Delta_L}{L}$, $\varepsilon_C = \dfrac{\mp \Delta_b}{b}$. Poisson's ratio: $\mu = \left[\dfrac{\varepsilon_c}{\varepsilon_L}\right]$. Hooke's law $\sigma = E\varepsilon$, $\varepsilon = \dfrac{\sigma}{E}$: $\Delta_L = \varepsilon_L L = \dfrac{\sigma}{E}L = \dfrac{P}{EA}L$, $\Delta_c = \varepsilon_c b = \dfrac{\mu \sigma}{E}b = \dfrac{\mu P}{EA}b$.
Temperature 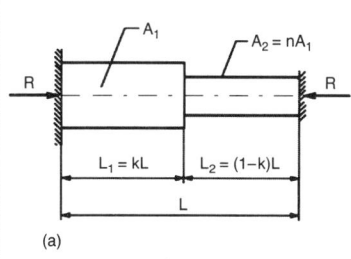 (a) 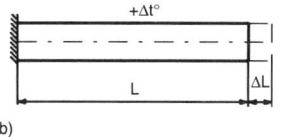 (b)	Case a: Reaction: $R = \dfrac{\alpha \cdot \Delta t^0 EA}{k + \dfrac{1-k}{n}}$, $n = \dfrac{A_2}{A_1}$, $k = \dfrac{L_1}{L}$. Axial force: $N = -R$ (compression). Stresses: $\sigma_1 = -\dfrac{R}{A_1} = -\dfrac{\alpha \cdot \Delta t^0 E}{k + \dfrac{1-k}{n}}$, $\sigma_2 = -\dfrac{R}{nA_1} = -\dfrac{\alpha \cdot \Delta t^0 E}{k(n-1)+1}$. For $A_1 = A_2$: $\sigma = \sigma_1 = \sigma_2 = -\alpha \cdot \Delta t^0 E$, $\Delta t^0 = T_o^0 - T_c^0$ Where T_o^0 and T_c^0 are original and considered temperatures. α = coefficient of linear expansion $\Delta t^0 > 0$ tension stress, $\Delta t^0 < 0$ compression stress. Case b: Deformation: $\Delta_L^t = \alpha \cdot \Delta t^0 L$.

6 Basis of Structural Analysis

─────────────── NOTES ───────────────

Example for Tables 1.2 and 1.3a. Bending

Given. Shape $W\,14 \times 30$, $L = 6$ m

Area $A = 8.85$ in^2 = $8.85 \times 2.54^2 = 57.097$ cm^2

Depth $h = 13.84$ in = $13.84 \times 2.54 = 35.154$ cm

Web thickness $d = 0.270$ in = $0.270 \times 2.54 = 0.686$ cm

Flange width $b = 6.730$ in = $6.730 \times 2.54 = 17.094$ cm

Flange thickness $t = 0.385$ in = $0.385 \times 2.54 = 0.978$ cm

Moment of inertia $I_z = 291$ in^4 = $291 \times 2.54^4 = 12{,}112.3$ cm^4

Section modulus $S = 42.0$ in^3 = $42.0 \times 2.54^3 = 688.26$ cm^3

Weight of the beam $\omega = 30$ lb/ft = $30 \times 4.448/0.3048 = 437.8$ N/m
$= 0.4378$ kN/m

Load $P = 80$ kN

Allowable stress (assumed) $[\sigma] = 196.2$ MPa, $[\tau] = 58.9$ MPa

Required. Compute σ_{max} and τ_{max}.

Solution.
$$M = \frac{\omega L^2}{8} + \frac{PL}{4} = \frac{0.4378 \times 6^2}{8} + \frac{80 \times 6}{4} = 121.97 \text{ kN}\cdot\text{m}$$

$$V = \frac{\omega L}{2} + \frac{P}{2} = \frac{0.4378 \times 6}{2} + \frac{80}{2} = 41.31 \text{ kN}$$

$$\sigma_{max} = \frac{M}{S} = \frac{121.97 \times 100 \text{ kN}\cdot\text{cm}}{688.26 \text{ cm}^3} = 17.72 \text{ kN/cm}^2 = 177{,}215.0 \text{ kN/m}^2$$

$$= 177.215 \text{ MPa} < 196.2 \text{ MPa}$$

$$\tau_{max} = \frac{V}{I_z d}\left[bt\left(\frac{h}{2} - \frac{t}{2}\right) + \frac{d\left(\frac{h}{2} - t\right)^2}{2}\right] = 1.890 \text{ kN/cm}^2 = 18{,}900 \text{ kN/m}^2$$

$$= 18.9 \text{ MPa} < 58.9 \text{ MPa}$$

TABLE 1.2 Stress and Strain: Bending

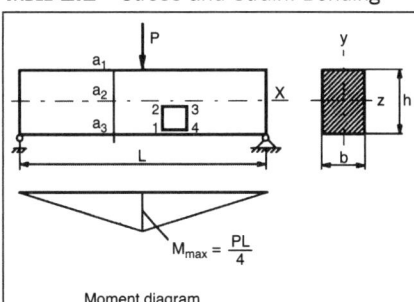

Moment diagram

Shear diagram

Stresses in two dimensions

Bending stress: $\sigma = \dfrac{M}{I_z} \cdot y$.

Shear stress: $\tau = \dfrac{VS}{I_z b}$.

Stresses in xy plane:
$$\sigma_y = 0, \quad \sigma_x = \sigma, \quad \tau_{xz} = \tau_{yz} = \tau.$$

Principal stresses:
$$\sigma_{\substack{max \\ min}} = \dfrac{\sigma}{2} \pm \dfrac{1}{2}\sqrt{\sigma^2 + 4\tau^2}.$$

Maximum shear (min) stresses:
$$\tau_{\substack{max \\ min}} = \pm \dfrac{1}{2}\sqrt{\sigma^2 + 4\tau^2}.$$

The principal stress and maximum (min) shear stresses lie at 45° to each other.

Stress diagrams

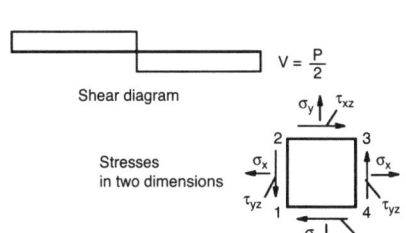

σ diagram: $\sigma_{a_1} = +\dfrac{M}{S}, \quad \sigma_{a_2} = 0, \quad \sigma_{a_3} = -\dfrac{M}{S}$.

τ diagram: $\tau_{a_1} = 0, \quad \tau_{a_2} = \dfrac{VS}{I_z b} = \dfrac{3V}{2A}, \quad \tau_{a_3} = 0$.

σ_{max} diagram:
$$\sigma_{a_1} = +\dfrac{M}{S}, \quad \sigma_{a_2} = +\tau = +\dfrac{3V}{2A}, \quad \sigma_{a_3} = 0.$$

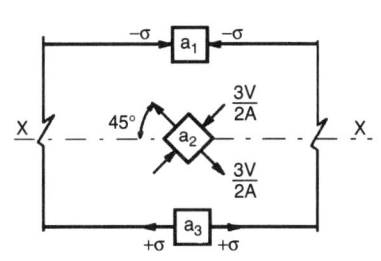

σ_{min} diagram:
$$\sigma_{a_1} = 0, \quad \sigma_{a_2} = -\tau = -\dfrac{3V}{2A}, \quad \sigma_{a_3} = -\dfrac{M}{S}.$$

τ_{max} diagram:
$$\tau_{a_1} = \tau_{a_3} = +\dfrac{\sigma}{2} = +\dfrac{M}{2S}, \quad \tau_{a_2} = +\tau = +\dfrac{3V}{2A}.$$

τ_{min} diagram:
$$\tau_{a_1} = \tau_{a_3} = -\dfrac{\sigma}{2} = -\dfrac{M}{2S}, \quad \tau_{a_2} = -\tau = -\dfrac{3V}{2A}.$$

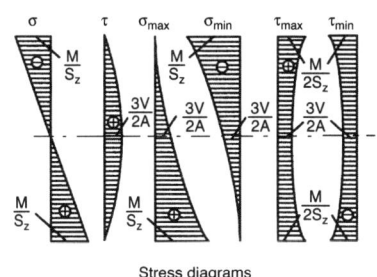

Stress diagrams

Note:

$+$ — Tension

$-$ — Compression

NOTES

Strain Energy

Stressing a bar stores energy in it. For an axial load P and a deformation e, the energy stored is

$$U = \frac{1}{2} Pe \tag{1.1a}$$

assuming the load is applied gradually and the bar is not stressed beyond the proportional limit. The equation represents the area under the load-deformation curve up to the load P.

Another useful equation for energy, in·lb, is

$$U = \frac{f^2}{2E} AL \tag{1.1b}$$

where f = unit stress, psi
E = modulus of elasticity of material, psi
A = cross-sectional area, in²
L = length of bar, in

Since AL is the volume of the bar, the term $f^2/2E$ gives the energy stored per unit of volume. It represents the area under the stress-strain curve up to the stress f.

Modulus of resilience is the energy stored per unit of volume in a bar stressed by a gradually applied axial load up to the proportional limit. This modulus is a measure of the capacity of the material to absorb energy without danger of being permanently deformed. It is important in designing members to resist energy loads.

Equation (1.1a) is a general equation that holds true when the **principle of superposition** applies (the total deformation produced at a point by a system of forces is equal to the sum of the deformations produced by each force). In the general sense, P in Eq. (1.2a) represents any group of statically interdependent forces that can be completely defined by one symbol, and e is the corresponding deformation.

The strain-energy equation can be written as a function of either the load or the deformation. For axial tension or compression, strain energy, in inch-pounds, is given by

$$U = \frac{P^2 L}{2AE} \qquad U = \frac{AEe^2}{2L} \tag{1.2a}$$

where P = axial load, lb
e = total elongation or shortening, in
L = length of member, in
A = cross-sectional area, in²
E = modulus of elasticity, psi

For pure shear:

$$U = \frac{V^2 L}{2AG} \qquad U = \frac{AGe^2}{2L} \tag{1.2b}$$

Continued on page 10

TABLE 1.3 Stress and Strain: Bending

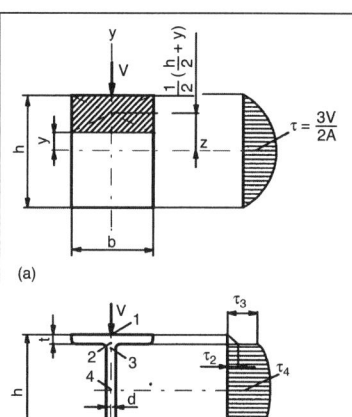

(a)

(b)

Bending in two directions

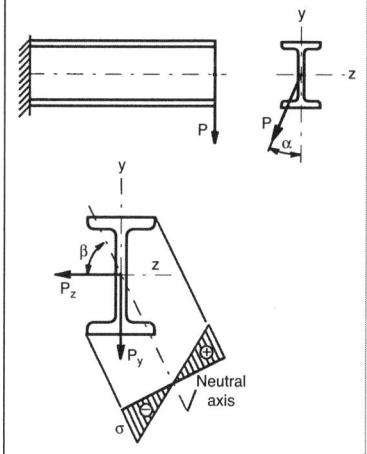

Shear stress: $\tau = \dfrac{VS}{I_z b}$

Case a: $S_y = \dfrac{b}{2}\left(\dfrac{h}{2} - y\right)\left(\dfrac{h}{2} + y\right) = \dfrac{b}{2}\left(\dfrac{h^2}{4} - y^2\right),$

$$\tau = \dfrac{V \cdot \dfrac{b}{2}\left(\dfrac{h^2}{4} - y^2\right)}{\dfrac{bh^3}{12} \cdot b} = \dfrac{6V}{bh^3}\left(\dfrac{h^2}{4} - y^2\right).$$

For $y = \pm\dfrac{h}{2}: \tau = 0,$ for $y = 0: \tau = \dfrac{3V}{2A}.$

Case b: $\tau_1 = 0,$

$$\tau_2 = \dfrac{V}{I_z b} bt\left(\dfrac{h}{2} - \dfrac{t}{2}\right), \quad \tau_3 = \dfrac{V}{I_z d} bt\left(\dfrac{h}{2} - \dfrac{t}{2}\right),$$

$$\tau_4 = \dfrac{V}{I_z d}\left[bt\left(\dfrac{h}{2} - \dfrac{t}{2}\right) + \dfrac{d\left(\dfrac{h}{2} - t\right)^2}{2}\right].$$

Bending moments.

Moment due to force P: $M = \sqrt{M_z^2 + M_y^2},$

$M_z = M\cos\alpha, \quad M_y = M\sin\alpha,$

$\left[\dfrac{M_y}{M_z}\right] = [\tan\alpha].$

For case shown: $M_z = P_y L \cos\alpha, \; M_y = P_z L \sin\alpha,$
$M = PL.$

$$\sigma = \pm M\left(\dfrac{y\cos\alpha}{I_z} + \dfrac{z\sin\alpha}{I_y}\right).$$

Stress:

$$\sigma_{max} = \pm \dfrac{M}{S_z}\left(\cos\alpha + \dfrac{S_z}{S_y}\sin\alpha\right).$$

Neutral axis: $\tan\beta = \dfrac{I_z}{I_y}\tan\alpha.$

Deflection in direction of force P: $\Delta = \sqrt{\Delta_z^2 + \Delta_y^2}.$

For case shown: $\Delta_z = \dfrac{P_z L^3}{3EI_y}, \quad \Delta_y = \dfrac{P_y L^3}{3EI_z}.$

where V = shearing load, lb
e = shearing deformation, in
L = length over which deformation takes place, in
A = shearing area, in^2
G = shearing modulus, psi

For torsion:

$$U = \frac{T^2 L}{2JG} \qquad U = \frac{JG\phi^2}{2L} \qquad (1.2c)$$

where T = torque, in·lb
ϕ = angle of twist, rad
L = length of shaft, in
J = polar moment of inertia of cross section, in^4
G = shearing modulus, psi

For pure bending (constant moment):

$$U = \frac{M^2 L}{2EI} \qquad U = \frac{EI\theta^2}{2L} \qquad (1.2d)$$

where M = bending moment, in·lb
θ = angle of rotation of one end of beam with respect to other, rad
L = length of beam, in
I = moment of inertia of cross section, in^4
E = modulus of elasticity, psi

For beams carrying transverse loads, the total strain energy is the sum of the energy for bending and that for shear.

Strain Energy in Structural Members*

Strain energy is generated in structural members when they are acted on by forces, moments, or deformations. Formulas for strain energy U, for shear, torsion, and bending in beams, columns, and other structural members are as follows:

Strain Energy in Shear

For a member subjected to pure shear, strain energy is given by

$$U = \frac{V^2 L}{2AG} \qquad (1.3)$$

$$U = \frac{AG\Delta^2}{2L} \qquad (1.4)$$

---

*Brockenbrough and Merritt, *Structural Steel Designer's Handbook*, McGraw-Hill.

Continued on page 12

TABLE 1.4 Stress and Strain: Combination of Compression (Tension) and Bending

Compression (tension) and bending 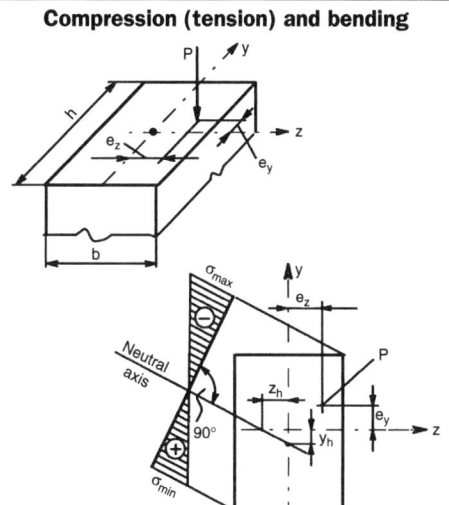	Stresses: $\sigma = \dfrac{P}{A} \pm \dfrac{M_y}{I_y} z \pm \dfrac{M_z}{I_z} y,$ $\sigma_{\max \atop \min} = \dfrac{P}{A} \pm \dfrac{M_y}{S_y} \pm \dfrac{M_z}{S_z},$ $M_y = P \cdot e_z, \quad M_z = P \cdot e_y,$ $I_y = \dfrac{h \cdot b^3}{12}, \quad I_z = \dfrac{b \cdot h^3}{12},$ $S_y = \dfrac{h \cdot b^2}{6}, \quad S_z = \dfrac{b \cdot h^2}{6}.$ Neutral axis: $y_n = \dfrac{i_z^2}{e_y}, \quad z_n = \dfrac{i_y^2}{e_z}.$ $i_z = \sqrt{I_z/A}, \quad i_y = \sqrt{I_y/A}, \quad A = b \cdot h.$
Buckling 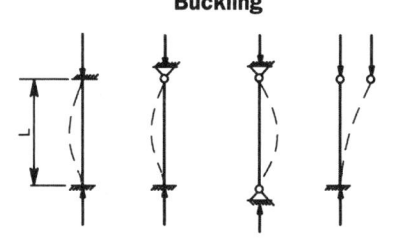 k = 0.5 | 0.7 | 1.0 | 2.0	Euler's formula: $P_e = \dfrac{\pi^2 EI}{(kL)^2} \quad \text{for} \quad \lambda_{\min} \geq \pi \sqrt{\dfrac{E}{R_e}},$ where R_e is the elastic buckling strength. $\lambda_{\min} = \dfrac{kL}{i_{\min}}, \quad \text{stress: } \sigma_{\max} \leq \dfrac{\pi^2 E}{\lambda_{\min}^2}.$
Axial compression (tension) and bending 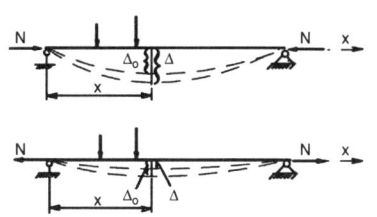	Stresses: Compression: $\sigma_{\max} = \dfrac{N}{A} + \dfrac{M_0}{S_z} + \dfrac{N}{S_z} \cdot \dfrac{\Delta_0}{1 - \dfrac{N}{P_e}},$ Tension: $\sigma_{\max} = \dfrac{N}{A} + \dfrac{M_0}{S_z} - \dfrac{N}{S_z} \cdot \dfrac{\Delta_0}{1 + \dfrac{N}{P_e}},$ where M_0 = max. moment and Δ_0 = max. deflection due to transverse loading.

--- NOTES ---

where V = shear load
Δ = shear deformation
L = length over which deformation takes place
A = shear area
G = shear modulus of elasticity

Strain Energy in Torsion

For a member subjected to torsion

$$U = \frac{T^2 L}{2JG} \tag{1.5}$$

$$U = \frac{JG\theta^2}{2L} \tag{1.6}$$

where T = torque
Δ = angle of twist
L = length over which the deformation takes place
J = polar moment of inertia
G = shear modulus of elasticity

Strain Energy in Bending

For a member subjected to pure bending (constant moment)

$$U = \frac{M^2 L}{2EI} \tag{1.7}$$

$$U = \frac{EI\theta^2}{2L} \tag{1.8}$$

where M = bending moment
θ = angle through which one end of beam rotates with respect to the other end
L = length over which the deformation takes place
I = moment of inertia
E = modulus of elasticity

For beams carrying transverse loads, the total strain energy is the the sum of the energy for bending and that for shear.

Example for Table 1.5. Torsion

Given. Cantilever beam, $L = 1.5$ m, for profile see Table 1.5c
$h = 70$ cm, $h_1 = 30$ cm, $h_2 = 60$ cm, $h_3 = 40$ cm, $b_1 = 4.5$ cm, $b_2 = 2.5$ cm, $b_3 = 5.5$ cm

Material: Steel, $G = 800$ kN/cm² $= 8000$ (MPa)

Torsion moment $M_t = 40$ kN·m

Required. Compute τ_{max} and ϕ^0.

Continued on page 14

Stress and Strain: Methods of Analysis 13

TABLE 1.5 Stress and Strain: Torsion

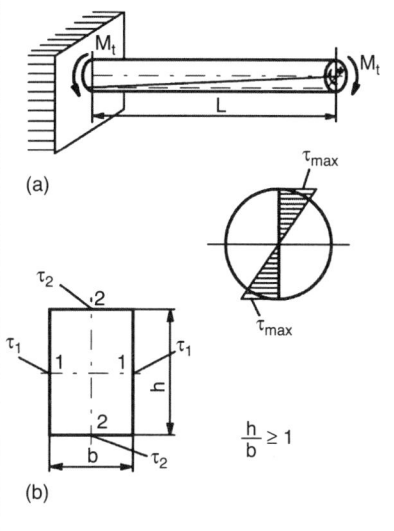

Bar of circular cross section

Stress: $\tau_{max} = \dfrac{M_t}{I_p} \cdot \dfrac{d}{2} = \dfrac{M_t}{S_p}$,

$I_p = \dfrac{\pi d^4}{32} \approx 0.1 d^4$, $S_p = \dfrac{\pi d^3}{16} \approx 0.2 d^3$.

Angle of twist: $\phi^0 = \dfrac{180^0}{\pi} \cdot \dfrac{M_t L}{G I_p}$

where G = shear modulus of elasticity.

Bar of rectangular cross section

Stress: $\tau_{max} = \dfrac{M_t}{S_t}$.

Angle of twist: $\phi^0 = \dfrac{180^\circ}{\pi} \cdot \dfrac{M_t L}{G I_t}$.

If $\dfrac{h}{b} > 10$: $I_t = \dfrac{h b^3}{3}$, $S_t = \dfrac{I_t}{b} = \dfrac{h b^2}{3}$.

If $\dfrac{h}{b} \leq 10$: $I_t = c_1 \cdot b^4$, $S_t = c_2 \cdot b^3$.

In point 1: $\tau_1 = \tau_{max}$, in point 2: $\tau_2 = c_3 \cdot \tau_{max}$.

$\dfrac{h}{b} =$	1.0	1.5	2.0	3.0	4.0	6.0	8.0	10.0	For
c_1	0.140	0.294	0.457	0.790	1.123	1.789	2.456	3.123	$\dfrac{h}{b} > 10$
c_2	0.208	0.346	0.493	0.801	1.150	1.789	2.456	3.123	
c_3	1.000	0.859	0.795	0.753	0.745	0.743	0.742	0.742	0.740

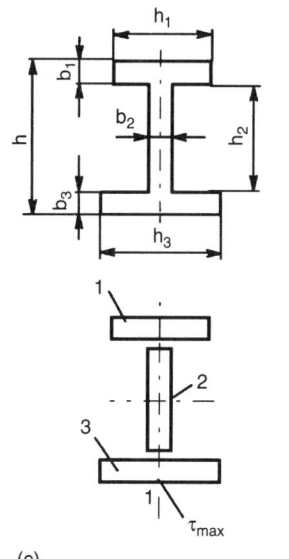

Profile consisting of rectangular cross sections

Geometric properties: $I_t = \sum_{i=1}^{i=n} I_{t_i}$, $S_t = \dfrac{I_t}{b_{max}}$,

$n = 3$.

Assumed: $\dfrac{h_1}{b_1} < 10$, $\dfrac{h_2}{b_2} > 10$, $\dfrac{h_3}{b_3} < 10$,

$b_3 > b_1$, $b_3 > b_2$ (i.e., $b_3 = b_{max}$)

$I_{t_1} = c_1 b_1^4$, $I_{t_2} = \dfrac{h_2 b_2^3}{3}$, $I_{t_3} = c_1 b_3^4$,

$I_t = I_{t_1} + I_{t_2} + I_{t_3}$, $S_t = \dfrac{I_t}{b_3}$.

Stress: $\tau_{max} = \dfrac{M_t}{S_t}$ (in point 1).

Angle of twist: $\phi^0 = \dfrac{180^\circ}{\pi} \cdot \dfrac{M_t L}{G I_t}$.

14   Basis of Structural Analysis

--- NOTES ---

Solution. $\dfrac{h_1}{b_1} = \dfrac{30}{4.5} = 6.67 < 10$, $c_1 = 2.012$,

$\dfrac{h_2}{b_2} = \dfrac{60}{2.5} = 24 > 10$, $\quad \dfrac{h_3}{b_3} = \dfrac{40}{5.5} = 7.27 < 10$, $c_1 = 2.212$

$I_{t_1} = c_1 b_1^4 = 2.012 \times 4.5^4 = 825.04 \text{ cm}^4$, $\quad I_{t_3} = c_1 b_3^4 = 2.212 \times 5.5^4 = 2024.12 \text{ cm}^4$

$I_{t_2} = \dfrac{h_2 b_2^3}{3} = \dfrac{60 \times 2.5^3}{3} = 312.5 \text{ cm}^4$, $\quad \sum I_t = I_{t_1} + I_{t_2} + I_{t_3} = 3161.66 \text{ cm}^4$

$$S_t = \dfrac{I_t}{b_{max}} = \dfrac{3161.66}{5.5} = 574.85 \text{ cm}^3,$$

$$\tau_{max} = \dfrac{40 \times 100}{574.85} = 6.958 \text{ kN/cm}^2 = 69{,}580 \text{ kN/m}^2 = 69.58 \text{ MPa}$$

$$\phi^\circ = \dfrac{180}{\pi} \cdot \dfrac{M_t L}{GI_t} = \dfrac{180}{3.14} \cdot \dfrac{40 \times 100 \times 1.5 \times 100}{800 \times 3161.66} = 13.6^\circ$$

Stress-Strain Relations

When a material is subjected to external forces, it develops one or more of the following types of strain: linear elastic, nonlinear elastic, viscoelastic, plastic, and anelastic. Many structural materials exhibit linear elastic strains under design loads. For these materials, unit strain is proportional to unit stress until a certain stress, the proportional limit, is exceeded (point A in Fig. 1.1a to c). This relationship is known as **Hooke's law**.

For axial tensile or compressive loading, this relationship may be written

$$f = E\varepsilon \quad \text{or} \quad \varepsilon = \dfrac{f}{E} \tag{1.9}$$

where f = unit stress
ε = unit strain
E = Young's modulus of elasticity

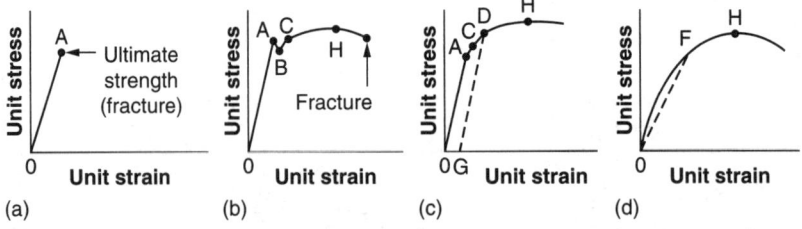

FIGURE 1.1   Relationship of unit stress and unit strain for various materials. (a) Brittle. (b) Linear elastic with a distinct proportional limit. (c) Linear elastic with an indistinct proportional limit. (d) Nonlinear.

Continued on page 16

TABLE 1.6 Stress and Strain: Curved Beams

Curved beam (transverse bending)

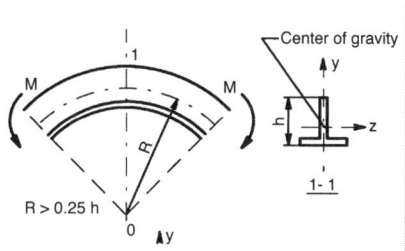

Stresses:
$$\sigma_y = \frac{M}{A \cdot c} \cdot \frac{y - R_0}{y}, \quad R_0 = \frac{\sum A_i}{\sum \frac{A_i}{R_i}}.$$

$$c = R - R_0$$

If $\frac{h}{R} \leq 0.5$, $c = \frac{I_z}{A \cdot R}$ for all cross section types.

For case shown:
$$A = A_1 + A_2, \quad R_0 = \frac{A_1 + A_2}{\frac{A_1}{R_1} + \frac{A_2}{R_2}},$$

$$\sigma_a = \frac{M}{A \cdot c} \cdot \frac{R_a - R_0}{R_a}, \quad \sigma_b = \frac{M}{A \cdot c} \cdot \frac{R_b - R_0}{R_b}.$$

$+\sigma$ — Tension
$-\sigma$ — Compression

Curved beam (axial force and bending)

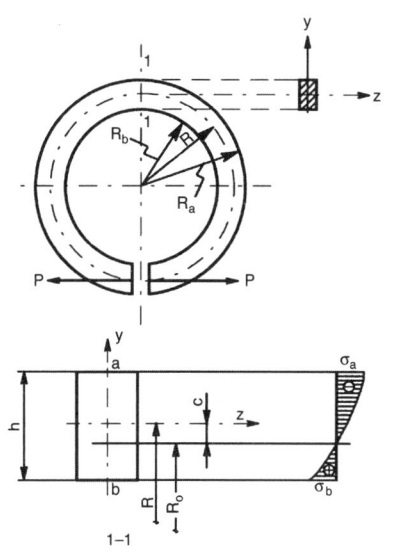

Stresses: $\sigma_\rho = \frac{N}{A} \pm \frac{M}{A \cdot c} \cdot \frac{\rho - R_0}{R_0}$.

For case shown: $c = R - R_0$,

$$R_0 = \frac{h}{\ln \frac{R_a}{R_b}} \quad \text{or} \quad R_0 \approx R\left[1 - \frac{1}{12}\left(\frac{h}{R}\right)^2\right].$$

$$N = P, \quad M = 2PR,$$

$$\sigma_a = \frac{P}{bh} - \frac{2PR}{bhc} \cdot \frac{R_a - R_0}{R_a},$$

$$\sigma_b = \frac{P}{bh} + \frac{2PR}{bhc} \cdot \frac{R_0 - R_b}{R_b}.$$

Note. For beams with circular cross section:

$$R_0 = \frac{1}{2}\left(R + \sqrt{R^2 - \frac{d^2}{R}}\right) \quad \text{or} \quad R_0 \approx R\left[1 - \frac{1}{16}\left(\frac{d}{R}\right)^2\right],$$

d = diameter of cross section.

16 Basis of Structural Analysis

NOTES

Within the elastic limit, there is no permanent residual deformation when the load is removed. Structural steels have this property.

In nonlinear elastic behavior, stress is not proportional to strain, but there is no permanent residual deformation when the load is removed. The relation between stress and strain may take the form

$$\varepsilon = \left(\frac{f}{K}\right)^n \tag{1.10}$$

where K = pseudoelastic modulus determined by test
n = constant determined by test

Viscoelastic behavior resembles linear elasticity. The major difference is that in linear elastic behavior, the strain stops increasing if the load does; but in viscoelastic behavior, the strain continues to increase although the load becomes constant and a residual strain remains when the load is removed. This is characteristic of many plastics.

Anelastic deformation is time-dependent and completely recoverable. Strain at any time is proportional to the change in stress. Behavior at any given instant depends on all prior stress changes. The combined effect of several stress changes is the sum of the effects of the several stress changes taken individually.

Example for Table 1.7. Continuous deep beam

Given. Beam $L = 3.0$ m, $h = 2.0$ m, $c = 0.3$ m, thickness $b = 0.3$ m, $w = 200$ kN/m

Required. Compute Z, D, d, d_0, and σ_{max} for center of span and support.

Solution. At center of span:

$Z = D = \alpha_z \times 0.5wL = 0.186 \times 0.5 \times 200 \times 3.0 = 55.8$ kN
$d = \alpha_d \times 0.5L = 0.888 \times 0.5 \times 3.0 = 1.33$ m
$d_0 = \alpha_{d_0} \times 0.5L = 0.124 \times 0.5 \times 3.0 = 0.19$ m
$\sigma_{max} = \alpha_\sigma \times w/b = 1.065 \times 200/0.3 = 710$ kN/m²
$\phantom{\sigma_{max}} = 0.71$ MPa (tension)

At center of support:

$Z = D = \alpha_z \times 0.5wL = 0.428 \times 0.5 \times 200 \times 3.0 = 128.4$ kN
$d = \alpha_d \times 0.5L = 0.656 \times 0.5 \times 3.0 = 0.984$ m
$d_0 = \alpha_{d_0} \times 0.5L = 0.036 \times 0.5 \times 3.0 = 0.05$ m
$\sigma_{max} = \alpha_\sigma \times w/b = -9.065 \times 200/0.3 = -6043.3$ kN/m²
$\phantom{\sigma_{max}} = -6.04$ MPa (compression)

Continued on page 18

Stress and Strain: Methods of Analysis

TABLE 1.7 Stress and Strain: Continuous Deep Beams

Formulas: Maximum tensile and compressive stresses $\sigma_{max} = \alpha_\sigma \cdot w$
Resultant tensile (Z) and compressive (D) forces $Z = D = \alpha_z \cdot 0.5wL$
$d = \alpha_d \cdot 0.5L$, $d_0 = \alpha_{d(0)} \cdot 0.5L$

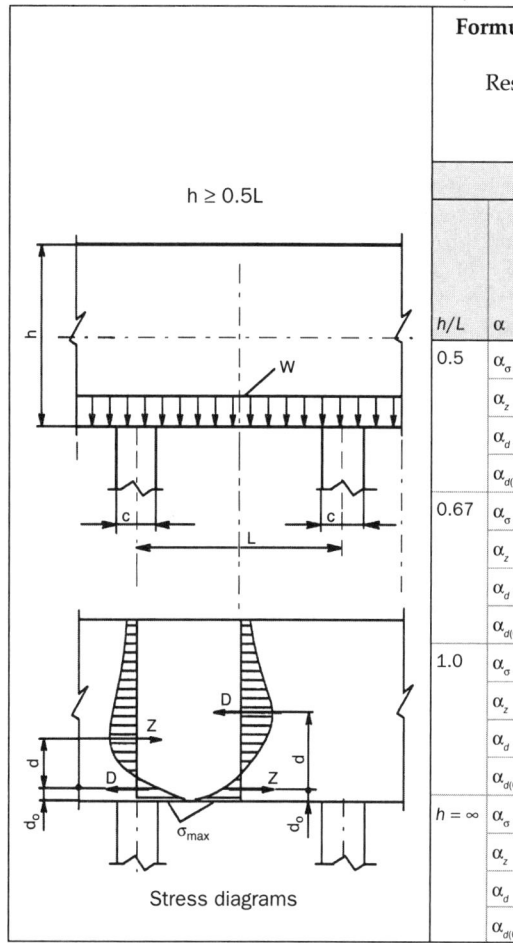

Stress diagrams

		At center of span			At center of support		
		c/L			c/L		
h/L	α	0.05	0.10	0.20	0.05	0.10	0.20
0.5	α_σ	1.317	1.313	1.289	−19.320	−9.317	−4.302
	α_z	0.240	0.239	0.235	0.515	0.485	0.375
	α_d	0.692	0.690	0.682	0.600	0.622	0.640
	$\alpha_{d(0)}$	0.129	0.128	0.127	0.022	0.039	0.062
0.67	α_σ	1.066	1.065	1.062	−19.066	−9.065	−4.062
	α_z	0.187	0.186	0.182	0.498	0.428	0.351
	α_d	0.890	0.888	0.880	0.620	0.656	0.686
	$\alpha_{d(0)}$	0.125	0.124	0.122	0.021	0.036	0.059
1.0	α_σ	1.002	1.002	1.002	−19.002	−9.002	−4.002
	α_z	0.178	0.177	0.172	0.497	0.424	0.324
	α_d	0.934	0.932	0.924	0.612	0.682	0.740
	$\alpha_{d(0)}$	0.124	0.123	0.121	0.021	0.036	0.059
$h = \infty$	α_σ	1.000	1.000	1.000	−19.000	−9.000	−4.000
	α_z	0.177	0.176	0.171	0.495	0.422	0.322
	α_d	0.938	0.936	0.930	0.612	0.674	0.746
	$\alpha_{d(0)}$	0.122	0.122	0.121	0.024	0.038	0.059

NOTES

NOTE: Tables 1.8 to 1.12 consider computation methods for elastic systems only.

Stress and Strain Failure Analysis

Material properties are usually determined from tests in which specimens are subjected to **simple stresses** under static or fluctuating loads. The attempt to apply these data to **bi- or triaxial stress fields** has resulted in the proposal of various theories of failure. Figure 1.2 shows the principal stresses on a triaxially stressed element. It is assumed, for simplicity, that $S_1 > S_2 > S_3$. Compressive stresses are negative.

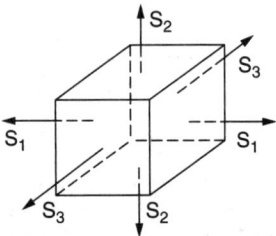

FIGURE 1.2 Principal stresses.

1. **Maximum stress theory** (Rankine) assumes failure occurs when the largest principal stress reaches the yield stress in a tension (or compression) specimen. That is $S_1 = \pm S_y$.

2. **Maximum shear theory** (Coulomb) assumes yielding (failure) occurs when the maximum shearing stress equals that in a simple tension (or compression) specimen at yield. Mathematically, $S_1 - S_3 = \pm S_y$.

3. **Maximum strain energy theory** (Beltrami) assumes failure occurs when the energy absorbed per unit volume equals the strain energy per unit volume in a tension (or compression) specimen at yield. Mathematically, $S_1^2 + S_2^2 + S_3^2 - 2\mu(S_1 S_2 + S_2 S_3 + S_3 S_1) = S_y^2$.

4. **Maximum distortion energy theory** (von Mises and Hencky) assumes yielding occurs when the distortion energy equals that in simple tension at yield. The distortion energy—that portion of the total energy which causes distortion rather than volume change—is

$$U_d = \frac{1+\mu}{3E}\left(S_1^2 + S_2^2 + S_3^2 - S_1 S_2 - S_2 S_3 - S_3 S_1\right)$$

Thus failure is defined by

$$S_1^2 + S_2^2 + S_3^2 - (S_1 S_2 + S_2 S_3 + S_3 S_1) = S_y^2$$

5. **Maximum strain theory** (Saint-Venant) claims failure occurs when the maximum strain equals the strain in simple tension at yield or $S_1 - \mu(S_2 + S_3) = S_y$.

6. **Internal friction theory** (Mohr). When the ultimate strengths in tension and compression are the same, this theory reduces to that of maximum shear. For principal stresses of opposite sign, failure is defined by $S_1 - \left(\frac{S_u}{S_{uc}}\right) S_2 = -S_{uc}$; if the signs are the same, $S_1 = S_u$ or $-S_{uc}$, where S_{uc} is the ultimate strength in compression.

Continued on page 20

TABLE 1.8 Stress and Strain: Dynamics, Transverse Oscillations of the Beams

Natural Oscillations of Systems with One Degree of Freedom

1 — Simple beam with one point mass

Forces:

P = weight of the load, mass: $m = \dfrac{P}{g}$

g = gravitational acceleration $\left(g = 981\,\dfrac{\text{cm}}{\text{s}^2}\right)$

P_i = force of inertia, $P_i = \mp ma$

a = acceleration

For shown beam:
Maximum bending moment:

$M_{max} = (P + P_i) \cdot \dfrac{a \cdot b}{L}$, Stress: $\sigma = \dfrac{M_{max}}{I_z} \cdot y$

Deflections:

Δ_{st} = static deflection due to load P

$\pm \Delta_i$ = max., min. deflection due to force P_i

$\Delta_{st(1)}$ = static deflection due to force $P = 1$

c = amplitude, $c = \pm \Delta_i$

Maximum shear for $a > b$

$V_{max} = (P + P_i) \cdot \dfrac{a}{L}$

Stress: $\tau = \dfrac{V_{max} \cdot S}{I_z \cdot t}$

2

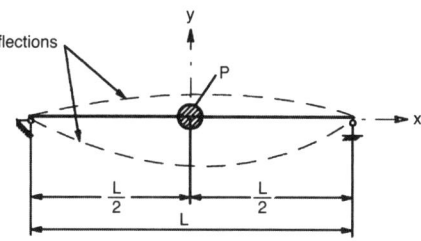

Force of inertia: $P_i = \dfrac{48cEI_z}{L^3}$

Maximum bending moment:

$M_{max} = \left(\dfrac{48cEI_z}{L^3} + P\right) \cdot \dfrac{L}{4}$

Maximum shear: $V_{max} = \dfrac{1}{2}\left(\dfrac{48cEI_z}{L^3} + P\right)$

3

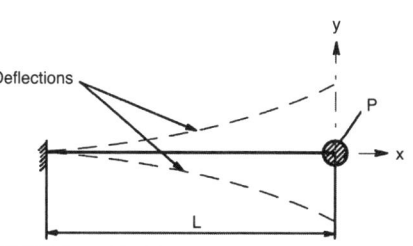

Force of inertia: $P_i = \dfrac{3cEI_z}{L^3}$

Maximum bending moment:

$M_{max} = \left(\dfrac{3cEI_z}{L^3} + P\right) \cdot L$

Maximum shear: $V_{max} = \dfrac{3cEI_z}{L^3} + P$

A **graphical representation** of the first four theories applied to a biaxial stress field is presented in Fig. 1.3. Stresses outside the bounding lines in the case of each theory mean failure (yield or fracture). A comparison with experimental data proves the distortion energy theory (4) best for ductile materials of equal tension-compression properties. When these properties are unequal, the internal energy theory (6) appears best. In practice, judging by some accepted codes, the maximum shear theory (2) is generally used for ductile materials and the maximum stress theory (1) for brittle materials.

Fatigue failures cannot be related, theoretically, to elastic strength and thus to the theories described. However, experimental results justify this, at least to a limited extent. Therefore, the theory evaluation given above holds for **fluctuating stresses**, provided that principal stresses at the maximum load are used and the **endurance strength** in simple bending is substituted for the yield strength.

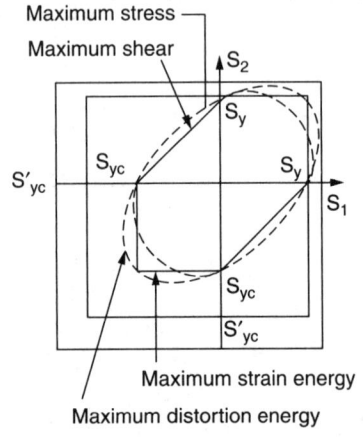

FIGURE 1.3 Biaxial stress field.

For example, a steel shaft, with 4-in diameter, is subjected to a bending moment of 120,000 in·lb, as well as a torque. If the yield strength in tension is 40,000 psi, what maximum torque can be applied under (*a*) the maximum shear theory and (*b*) the distortion energy theory?

$$S_x = \frac{Mc}{I} = \frac{120,000 \times 2}{12.55} = 19,100 \, \text{psi} \qquad S_{xy} = \frac{TC}{J} = \frac{T \times 2}{25.1} = 0.0798T$$

and

$$S_{M \cdot m} = \frac{S_x}{2} \pm \sqrt{\left(\frac{S_x}{2}\right)^2 + S_{xy}^2}$$

$$S_M - S_m = S_y \qquad \text{or} \qquad 2\sqrt{\left(\frac{19,100}{2}\right)^2 + (0.0798T)^2} = (40,000)^2 \qquad (a)$$

or

$$T = 221,000 \, \text{in·lb}$$

$$S_M^2 + S_m^2 - S_M S_m = S_y^2 \qquad (b)$$

Substituting and simplifying,

$$(9550)^2 + 3\sqrt{\left(\frac{19,100}{2}\right)^2 + (0.0798T)^2} = (40,000)^2$$

or

$$T = 255,000 \, \text{in·lb}$$

TABLE 1.9 Stress and Strain: Dynamics, Transverse Oscillations of the Beams

Diagram of continuous oscillations
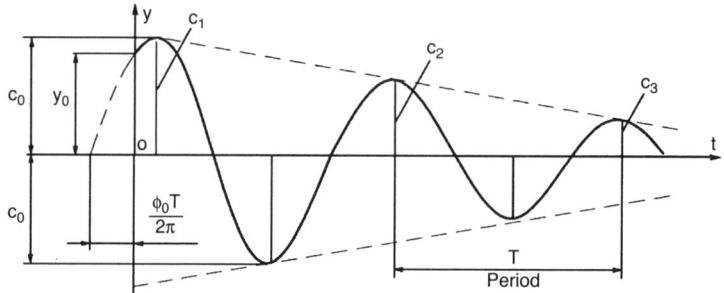

Equation of free continuous oscillations: $y = c \sin(\omega t + \phi_0)$

Where ϕ_0 = initial phase of oscillation, $\phi_0 = \arcsin\left(\dfrac{y_0}{c}\right)$

c_0 = amplitude, t = time, T = period of free oscillation, $T = \dfrac{2\pi}{\omega} = 2\pi\sqrt{\dfrac{\Delta_{st}}{g}}$

ω = frequency of natural oscillation, $\omega = \sqrt{\dfrac{g}{\Delta_{st}}}$

Diagram of damped oscillations

Equation of free damped oscillations: $y = c_0 e^{-kt/2m} \cdot \sin(\omega t + \phi_0)$

where c_0 = initial amplitude of oscillation, $c_0 = \sqrt{y_0^2 + \left(\dfrac{v_0 + y_0 k \cdot 2m}{\omega}\right)^2}$

ϕ_0 = initial phase of oscillation, $\phi_0 = \arcsin\left(\dfrac{y_0}{c_0}\right)$

y_0 = initial deflection

v_0 = beginner velocity of mass

e = logarithmic base, $e = 2.71828$

k = coefficient set according to material, mass, and rigidity

T = period of free oscillations, $T = 2\pi/\omega$

ω = frequency of free oscillation, $\omega = \sqrt{r/m - [k/2m]^2}$, For simple beam: $r = \dfrac{48EI_z}{L^3}$

22 Basis of Structural Analysis

NOTES

Spring type	Spring deflection	Spring force and bending stresses	
(a)	$F_1 = \dfrac{KPr^3}{3EI}(m+\beta)^3$ where $\alpha = \beta$ for finding K	When $\alpha = 0°$ to $90°$ $P = \dfrac{S\sigma}{u+\sin\beta}$ $\sigma = \dfrac{Pr(m+\sin\beta)}{S}$	When $\alpha = 90°$ to $180°$ $P = \dfrac{S\sigma}{u+r}$ $\sigma = \dfrac{Pr(m+1)}{S}$
(b)	$F_2 = \dfrac{2KPr^3}{3EI}\left(m+\dfrac{\beta}{2}\right)^3$ where $\alpha = \dfrac{\beta}{2}$ for finding K	$P = \dfrac{S\sigma}{L}$	
(c)	$F_3 = 2F_2 = \dfrac{4KPr^3}{3EI}\left(m+\dfrac{\beta}{2}\right)^3$ where $\alpha = \dfrac{\beta}{2}$ for finding K	$\sigma = \dfrac{PL}{S}$ Typical curved spring. (*Product Engineering*.)	
(d), (e)	$F_4 = F_5 = \dfrac{P}{3EI} \times \left[2Kr^3\left(m+\dfrac{\beta}{2}\right)^3 + (v-u)^3\right]$ where $\alpha = \dfrac{\beta}{2}$ for finding K	$P = \dfrac{S\sigma}{\lambda} = \dfrac{P\lambda}{S}$	

First Condition	Second Condition	λ
$u \geq v$	—	$u+r$
$u < v$	$(u-v) < (u+r)$	$u+r$
$u < v$	$(u-v) > (u+r)$	$v-u$
$u = 0$	$v \leq r$	r
$u = 0$	$v > r$	v

FIGURE 1.4 Deflection, force, and stress relations for curved springs. (*Product Engineering*.)

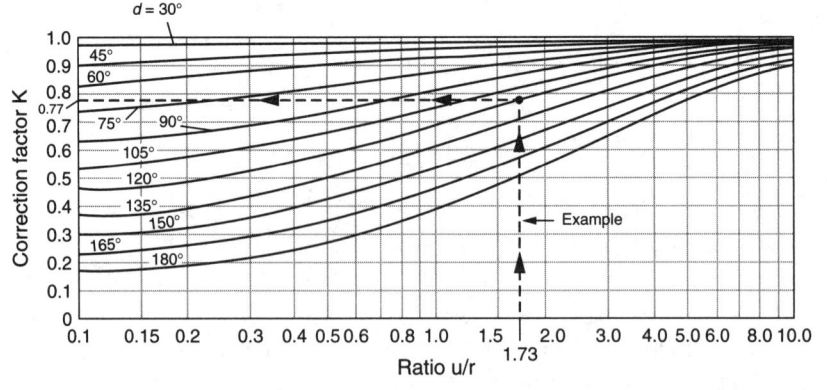

FIGURE 1.5 Correction factors for curved springs. (*Product Engineering*.)

TABLE 1.10 Stress and Strain: Dynamics, Transverse Oscillations of the Beams

Forced Oscillations of the Beams with One Degree of Freedom

Simple beam with one point mass

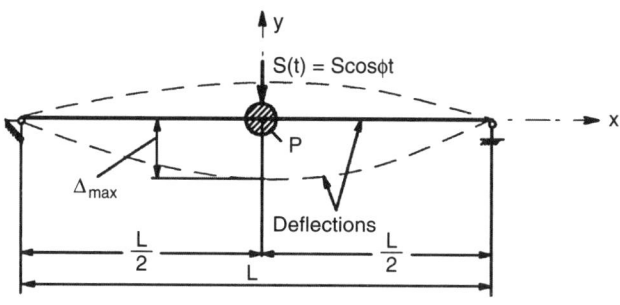

Forces:

P = weight of the load,

mass: $m = \dfrac{P}{g}\left(g = 981\dfrac{cm}{s^2}\right)$

$S(t)$ = vibrating force, assumed: $S(t) = S\cos\phi t$

P_i = force of inertia, $P_i = \dfrac{\Delta_{max} - \Delta_{st}}{\Delta st_1} - S\cos\phi t$

ϕ = frequency of force $S(t)$

$\Delta_{st(1)}$ = static deflection due to load $P = 1$

Deflections:

$\Delta_{max} = \Delta_{st(p)} + \Delta_{st(s)} + \Delta_i$

$\Delta_{st(p)}$ = static deflection due to load P

$\Delta_{st(s)}$ = static deflection due to force S

Δ_i = static deflection due to P_i

$\Delta_i = P_i \cdot \Delta_{st(1)}$

Equation of forced oscillations: $y = c \cdot e^{-kt/2m} \cdot \sin(\omega t + \phi_0) + \dfrac{g \cdot S(t)}{P(\omega^2 - \phi^2)} \cdot \cos\phi t$

$c \cdot e^{-kt/2m} \cdot \sin(\omega t + \phi_0)$ = free oscillation, $\quad \dfrac{g \cdot S(t)}{P(\omega^2 - \phi^2)} \cdot \cos\phi t$ = forced oscillation

ϕ_0 = beginner phase of oscillation, $\quad \phi_0 = \arcsin\left(\dfrac{y_0}{c_0}\right)$, $\quad y_0$ = beginner deflection

c_0 = amplitude of free oscillation, $\quad c_0 = c$,

c = amplitude of forced oscillation, $\quad c = k_D \cdot \Delta_{st(s)}$

k = coefficient set according to material, mass, and rigidity

ω = frequency of natural oscillation, $\quad T$ = period of oscillations, $\quad T = 2\pi/\omega$

k_D = dynamic coefficient, $k_D = \dfrac{1}{\sqrt{\left(1 - \dfrac{\phi^2}{\omega^2}\right)^2 + \left[\dfrac{k \cdot \phi}{m \cdot \omega^2}\right]^2}}$

If $k = 0$ (damped oscillation is not included): $k_D = \dfrac{1}{1 - \dfrac{\phi^2}{\omega^2}}$

e = logarithmic base, $\quad e = 2.71828$, $\quad g$ = gravitational acceleration $\left(g = 981\dfrac{cm}{s^2}\right)$

24 Basis of Structural Analysis

─────────── N O T E S ───────────

Example for Table 1.11. Bending
Given. Beam $W12 \times 65$, steel, $L = 3.0$ m
  Moment of inertia $I_z = 533 \text{ in}^4 \times 2.54^4 = 22{,}185 \text{ cm}^4$
  Section modulus $S = 87.9 \text{ in}^3 = 87.9 \times 2.54^3 = 1440.4 \text{ cm}^3$
  Modulus of elasticity $E = 29{,}000 \text{ kip/in}^2 = \dfrac{29{,}000 \times 4.48222}{2.54^2}$
  $= 20{,}147.6 \text{ kN/cm}^2$
  Weight of beam (concentrated load):
  $W = 65 \text{ lb/ft} \times 3.0 = 195 \times 4.448/0.3048 = 2845.7 \text{ N} = 2.8457 \text{ kN}$
  Load $P = 20$ kN, $h = 5$ cm

Required. Compute dynamic stress σ.

Solution. $\Delta_{st} = \dfrac{PL^3}{48EI_z} = \dfrac{20 \times (3 \times 100)^3}{48 \times 20{,}147.6 \times 22{,}185} = 0.025$ cm

$$k_D = 1 + \sqrt{1 + \dfrac{2h}{\Delta_{st}\left(1 + \beta\dfrac{W}{P}\right)}} = 1 + \sqrt{1 + \dfrac{2 \times 5}{0.025\left(1 + \dfrac{17}{35} \times \dfrac{2.8457}{20}\right)}} = 1 + 19.4 = 20.4$$

Bending moment $M_D = \dfrac{PL}{4} \cdot k_D = \dfrac{20 \times 3}{4} \times 20.4 = 306 \text{ kN} \cdot \text{m}$

Stress $\sigma = \dfrac{M_D}{S} = \dfrac{306 \times 100}{1440.4} = 21.24 \text{ kN/cm}^2 = 212{,}400 \text{ kN/m}^2 = 212.4$ MPa

---

Example for Table 1.11. Crane cable
Given. Load $P = 40$ kN, velocity $v = 5$ m/s
  Cable: diameter $d = 5.0$ cm, $A = 19.625 \text{ cm}^2$, $L = 30$ m,
  Modulus of elasticity $E = 29{,}000 \text{ kip/in}^2 = \dfrac{29{,}000 \times 4.48222}{2.54^2} = 20{,}147.6 \text{ kN/cm}^2$

Required. Compute dynamic stress σ for sudden dead stop.
Solution.

$$\Delta_{st} = \dfrac{PL}{EA} = \dfrac{40 \times 30 \times (100)}{20{,}147.6 \times 19.625} = 0.303 \text{ cm,}$$

$$k_D = \dfrac{v}{\sqrt{g \cdot \Delta_{st}}} = \dfrac{5 \times (100)}{\sqrt{981 \times (100) \times 0.303}} = 2.9$$

Stress:

$$\sigma = \dfrac{P}{A}(1 + k_D) = \dfrac{40}{19.625}(1 + 2.9) = 7.949 \text{ kN/cm}^2 = 79{,}490 \text{ kN/m}^2 = 79.45 \text{ MPa}$$

TABLE 1.11 Stress and Strain: Dynamics, Impact

Elastic Design	
Axial compression 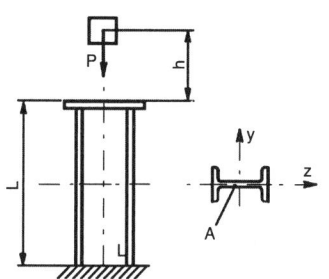	Dynamic coefficient: $$k_D = 1 + \sqrt{1 + \frac{v^2}{g\Delta_{st}\left(1+\beta\frac{W}{P}\right)}}$$ $$= 1 + \sqrt{1 + \frac{2h}{\Delta_{st}\left(1+\beta\frac{W}{P}\right)}}$$ where v = striking velocity, $v = \sqrt{2gh}$ g = earth's acceleration, $g = 9.81$ m/sec² Δ_{st} = deflection resulting from static load P W = weight of the structure β = coefficient for uniform mass
	For shown column: $\Delta_{st} = \dfrac{PL}{EA}$, $\beta = \dfrac{1}{3}$. Dynamic stress: $\sigma = -\dfrac{P}{A} \cdot k_D$
Bending 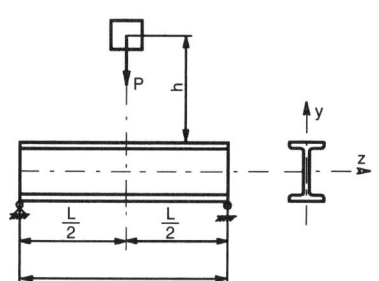	For shown beam: $\Delta_{st} = \dfrac{PL^3}{48EI_z}$, $\beta = \dfrac{17}{35}$. Dynamic bending moment: $M_D = \dfrac{PL}{4} \cdot k_D$ Dynamic shear: $V_D = \dfrac{P}{2} \cdot k_D$ For stresses see Table 1.3.
Crane cable 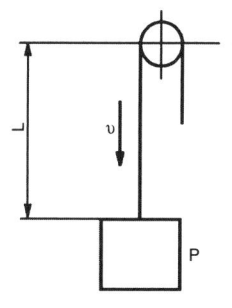	Sudden dead stop when the load P is going down. Dynamic coefficient: $$k_D = \frac{v}{\sqrt{g \cdot \Delta_{st}}}$$ where v = descent's velocity $$\Delta_{st} = \frac{PL}{EA}$$ Maximum stress in the cable: $$\sigma = \frac{P}{A}(1+k_D)$$ A = area of cable cross section

Analysis and Design of Flat Metal Springs

Type of spring	W_1 safe load[2]	F_1 deflection[2]
Flat parallel spring	$\dfrac{Sbt^2}{6l}$	$\dfrac{4Wl^3}{Ebt^3} = \dfrac{2}{3}\dfrac{Sl^2}{Et}$
Flat triangular spring	$\dfrac{Sbt^2}{6l}$	$\dfrac{6Wl^3}{Ebt^3} = \dfrac{Sl^2}{Et}$
Leaf spring	$\dfrac{SNbt^2}{6l}$ Where N = No. of leaves	$\dfrac{6Wl^3}{ENbt^3} = \dfrac{Sl^2}{Et}$ Where N = No. of leaves

FIGURE 1.6 Flat metal spring formulas.

W = save load or pull, lb (N)
F = deflection at point of application, in (cm)
S = safe tensile stress of material, lb/in² (kPa)
E = modulus of elasticity, 30×10^6 for steel (kPa)

TABLE 1.12 Stress and Strain: Dynamics, Impact

Elastic Design	
Column with buffer spring 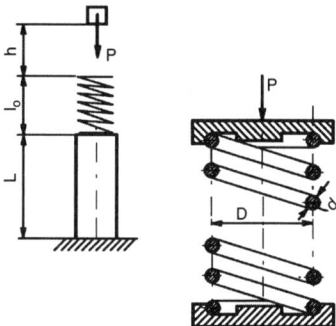	Cylindrical helical spring: D = average diameter d = spring wire's diameter n = number of effective rings G = shear modulus of elasticity for spring wire Dynamic coefficient: $$k_D = 1 + \sqrt{1 + \frac{2h}{P\left(\dfrac{8D^3 n}{Gd^4} + \dfrac{L}{EA}\right)}}$$ Dynamic stress: $\sigma = -\dfrac{P}{A} \cdot k_D$ (compression) E = modulus of elasticity for column A = area of column cross section
Motor mounted on the beam 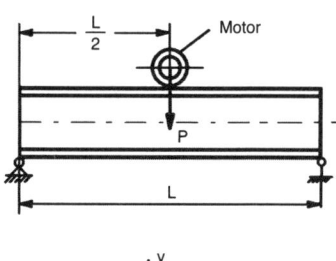 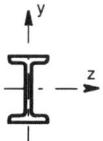 P = motor's weight, F_c = centrifugal force causing vertical vibration of the beam, $F_c = m\phi^2 r$ m = mass of rotative motor part r = radius of rotation n = revolutions per minute	Dynamic coefficient: $k_D = \dfrac{1}{1 - \dfrac{\phi^2}{\omega^2}}$, ϕ = frequency of force F_c, $\phi = \dfrac{n}{60} \cdot 2\pi = \dfrac{\pi n}{30}\left(\dfrac{1}{\text{s}}\right)$ ω = beam's free vibration frequency, $$\omega = \sqrt{\dfrac{g}{P\Delta}}\left(\dfrac{1}{\text{s}}\right)$$ Δ = beam's deflection by force $P = 1$ at the point of motor attachment $\left(\text{For shown case: } \Delta = \dfrac{L^3}{48EI_z}\right)$ Resonance: $\phi = \omega$, $n = \dfrac{30\phi}{\pi}$. Stresses: Static stress: $\sigma = \dfrac{PL}{4S_z}$, dynamic stress: $\sigma = \dfrac{F_c k_D L}{4S_z}$, $\sum\sigma = \dfrac{L}{4S_z}(P + F_c k_D)$.

28 Basis of Structural Analysis

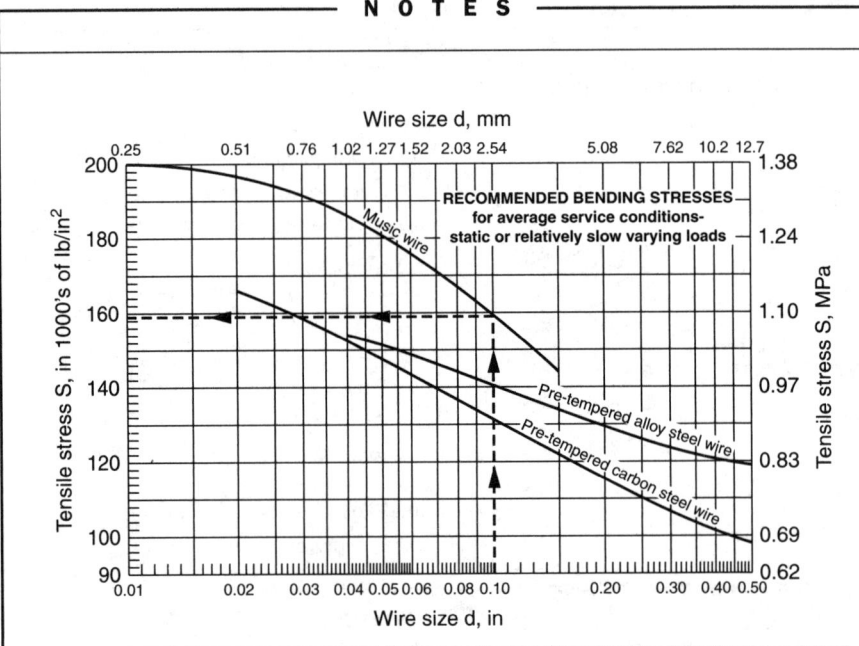

FIGURE 1.7 Recommended bending stresses for torsion springs. (*Product Engineering.*)

CHAPTER 2
Properties of Geometric Sections

─────── NOTES ───────

Properties of Geometric Sections of Columns

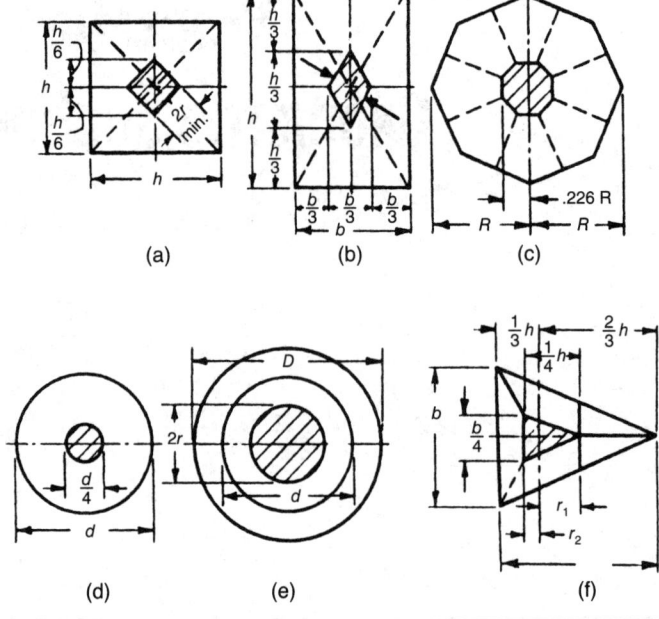

FIGURE 2.1   Column characteristics.

The *kern* is the area around the center of gravity of a cross section within which any load applied produces stress of only one sign throughout the entire cross section. Outside the kern, a load produces stresses of different sign. Figure 2.1 shows kerns (shaded) for various sections.

For a *circular ring*, the radius of the kern $r = D[1 + (d/D)^2]/8$.

For a *hollow square* (H and h = lengths of outer and inner sides), the kern is a square similar to Fig. 2.1a, where

$$r_{min} = \frac{H}{6}\frac{1}{\sqrt{2}}\left[1 + \left(\frac{h}{H}\right)^2\right] = 0.1179H\left[1 = \left(\frac{h}{H}\right)^2\right]$$

For a *hollow octagon*, R_a and R_i are the radii of circles circumscribing the outer and inner sides, respectively; thickness of wall = $0.9239(R_a - R_i)$; and the kern is an octagon similar to Fig. 2.1c, where $0.2256R$ becomes $0.2256R_a \times [1 + (R_i/R_a)^2]$.

Properties of Geometric Sections

TABLE 2.1 Properties of Geometric Sections: Tension, Compression, and Bending Structures

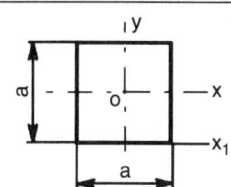

1. SQUARE

$$A = a^2, \quad I_x = I_y = \frac{a^4}{12}, \quad I_{x_1} = \frac{a^4}{3},$$

$$S_x = S_y = \frac{a^2}{6}, \quad r_x = r_y = \frac{a}{\sqrt{12}} = 0.289a, \quad Z = \frac{a^3}{4}.$$

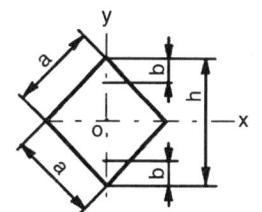

2. SQUARE
Axis of moments on diagonal

$$A = a^2, \quad h = a\sqrt{2} = 1.42a, \quad I_x = I_y = \frac{a^4}{12}, \quad S_x = S_y = \frac{a^3}{6\sqrt{2}} = 0.118a^3,$$

$$r_x = r_y = \frac{a}{\sqrt{12}} = 0.289a, \quad Z = \frac{a}{3\sqrt{2}} = 0.236a.$$

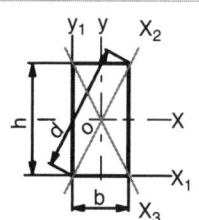

3. RECTANGLE

$$A = bh, \quad I_x = \frac{bh^3}{12}, \quad I_y = \frac{b^3 h}{12}, \quad I_{x_1} = \frac{bh^3}{3}, \quad I_{y_1} = \frac{b^3 h}{3},$$

$$S_x = \frac{bh^2}{6}, \quad S_y = \frac{b^2 h}{6}, \quad r_x = 0.289h, \quad r_y = 0.289b,$$

$$I_{x_2} = I_{x_3} = \frac{d^4 \sin \alpha}{48}.$$

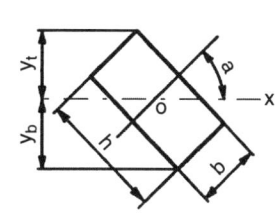

4. RECTANGLE
Axis of moments on any line through center of gravity

$$A = bh, \quad y_t = y_b = \frac{1}{2}(h \cos\alpha + b \sin\alpha),$$

$$I_x = \frac{bh}{12}(h^2 \cos^2\alpha + b^2 \sin^2\alpha), \quad S_x = \frac{bh(h^2 \cos^2\alpha + b^2 \sin^2\alpha)}{6(h \cos\alpha + b \sin\alpha)},$$

$$r_x = 0.289\sqrt{(h^2\cos^2 + b^2\sin^2 \alpha)}.$$

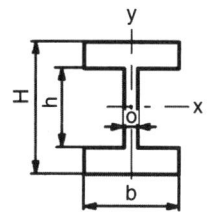

5. SYMMETRICAL SHAPE

$$A = ah + b(H - h),$$

$$I_x = \frac{ah^3}{12} + \frac{b}{12}(H^3 - h^3), \quad I_y = \frac{a^3 h}{12} + \frac{b^3}{12}(H - h),$$

$$S_x = \frac{b}{6H}(H^3 - h^3) + \frac{ah^3}{6H}, \quad S_y = \frac{a^3 h}{6b} + \frac{b^2}{6}(H - h).$$

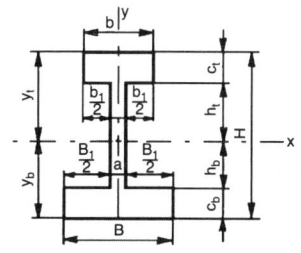

6. NONSYMMETRICAL SHAPE

$$A = bc_t + a(h_b + h_t) + Bc_b, \quad b_1 = b - a, \quad B_1 = B - a,$$

$$y_b = \frac{aH^2 + B_1 c_b^2 + b_1 c_t(2H - c_t)}{2(aH + B_1 c_b + b_1 c_t)}, \quad y_t = H - y_b,$$

$$I_x = \frac{1}{3}(By_b^3 - B_1 h_b^3 + by_t^3 - b_1 h_t^3).$$

NOTES

TABLE 2.1N Properties of Various Cross Sections

I = moment of inertia; I/c = section modulus; $r = \sqrt{I/A}$ = radius of gyration

Section	Moment of Inertia	Section Modulus	Radius of Gyration
Rectangle (centroidal)	$I = \dfrac{bh^3}{12}$ $\dfrac{I}{c} = \dfrac{bh^2}{6}$ $r = \dfrac{h}{\sqrt{12}} = 0.289h$	$\dfrac{bh^3}{3}$ $\dfrac{bh^2}{3}$ $\dfrac{h}{\sqrt{3}} = 0.577h$	$\dfrac{b^3h^3}{6(b^2+h^2)}$ $\dfrac{b^2h^2}{6\sqrt{b^2+h^2}}$ $\dfrac{bh}{\sqrt{6(b^2+h^2)}}$	$\dfrac{bh}{12}(h^2\cos^2 a + b^2\sin^2 a)$ $\dfrac{bh}{6}\left(\dfrac{h^2\cos^2 a + b^2\sin^2 a}{h\cos a + b\sin a}\right)$ $\sqrt{\dfrac{h^2\cos^2 a + b^2\sin^2 a}{12}}$
Hollow sections	$I = \dfrac{b}{12}(H^3 - h^3)$ $\dfrac{I}{c} = \dfrac{b}{6}\dfrac{H^3 - h^3}{H}$ $r = \sqrt{\dfrac{H^3 - h^3}{12(H-h)}}$	$\dfrac{H^4 - h^4}{12}$ $\dfrac{1}{6}\dfrac{H^4 - h^4}{H}$ $\sqrt{\dfrac{H^2 + h^2}{12}}$	$\dfrac{H^4 - h^4}{12}$ $\dfrac{\sqrt{2}}{12}\dfrac{H^4 - h^4}{H}$ $\sqrt{\dfrac{H^2 + h^2}{12}}$	$\dfrac{bh^3}{36} \; ; c = \dfrac{2}{3}h$ $\dfrac{bh^2}{24}$ $\dfrac{h}{\sqrt{18}}$
Triangle / Hexagon / Octagon	$I = \dfrac{bh^3}{12}$ $\dfrac{I}{c} = \dfrac{bh^2}{12}$ $r = \dfrac{h}{\sqrt{6}}$	$\dfrac{5\sqrt{3}}{16}R^4$ $\dfrac{5}{8}R^3$ $\sqrt{\dfrac{5}{24}}R$	$\dfrac{5\sqrt{3}}{16}R^4$ $\dfrac{5\sqrt{3}}{16}R^3$	$\dfrac{1+2\sqrt{2}}{6}R^4$ $0.6906R^3$ $0.475R$

Properties of Geometric Sections   33

TABLE 2.2 Properties of Geometric Sections: Tension, Compression, and Bending Structures

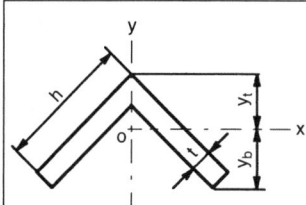

7. ANGLE with equal legs

$$A = t(2h-t), \quad y_t = \frac{h^2 + ht + t^2}{2(2h-t)\cos 45°}, \quad y_b = \frac{h+t-2c}{\sqrt{2}},$$

$$I_x = \frac{1}{3}\left[2c^4 - 2(c-t)^4 + t(h-2c+\frac{1}{2}t)^3\right],$$

$$c = y_t \cos 45°.$$

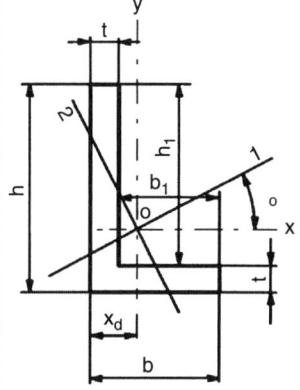

8. ANGLE with unequal legs

$$A = t(b+h_1) = t(h+b_1), \quad x_d = \frac{b^2 + h_1 t}{2(b+h_1)}, \quad y_d = \frac{h^2 + b_1 t}{2(h+b_1)},$$

$$I_x = \frac{1}{3}\left[t(h-y_d)^3 + by_d^3 - b_1(y_d - t)^3\right],$$

$$I_y = \frac{1}{3}\left[t(b-x_d)^3 + hx_d^3 - h_1(x_d - t)^3\right],$$

$$I_1 = I_{max} \quad \text{and} \quad I_2 = I_{min}, \quad \tan 2\varphi_0 = \frac{2I_{xy}}{I_y - I_x},$$

I_{xy} = product of inertia about axes x and y, $\quad I_{xy} = \pm \frac{bb_1 hh_1 t}{4(b+h_1)},$

$$I_{1(2)} = I_{max\,(min)} = \frac{1}{2}(I_y + I_x) \pm \frac{1}{2}\sqrt{(I_y - I_x)^2 + 4I_{xy}^2}.$$

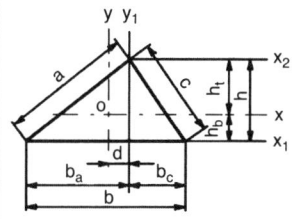

9. TRIANGLE

$$A = \frac{1}{2}bh, \quad h_b = \frac{1}{3}h, \quad h_t = \frac{2}{3}h, \quad d = \frac{1}{3}(b_a - b_c),$$

$$I_x = \frac{bh^3}{36}, \quad I_{x_1} = \frac{bh^3}{12}, \quad I_{x_2} = \frac{bh^3}{4},$$

$$I_y = \frac{hb(b^2 - b_a b_c)}{36}, \quad I_{y_1} = \frac{h(b_a^3 + b_c^3)}{12},$$

$$S_{x(b)} = \frac{bh^2}{12} \text{(for base)}, \quad S_{x(t)} = \frac{bh^2}{24} \text{(for point } A\text{)}, \quad r_x = \frac{h}{3\sqrt{2}} = 0.236h.$$

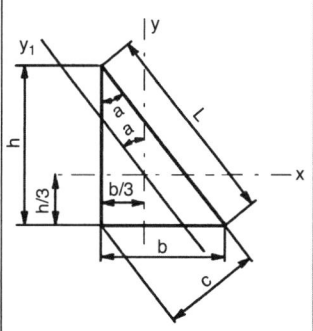

10. RECTANGULAR TRIANGLE

$$A = \frac{bh}{2} = \frac{cL}{2}, \quad I_x = \frac{bh^3}{36}, \quad I_y = \frac{hb^3}{36},$$

$$I_{y_1} = \frac{b^3 h^3}{36L^2} = \frac{Lc^3}{36},$$

or $\quad I_{y_1} = I_y \cos^2\alpha + I_x \sin^2\alpha + 2I_{xy} \sin\alpha \cos\alpha,$

$$\sin\alpha = \frac{b}{L}, \quad \cos\alpha = \frac{h}{L}, \quad I_{xy} = -\frac{b^2 h^2}{72},$$

$$r_x = \frac{h}{3\sqrt{2}} = 0.236h.$$

NOTES

TABLE 2.2N Torsion of Shafts of Various Cross Sections
G = Shear modulus of elasticity, psi

Cross Section	Torsional Resisting Moment M_t	Angular Deflection a_l (Length = 1 in, Radius = 1 in)		Work of Torsion (V = Volume)
		In Terms of Torsional Moment	In Terms of Max Shear	
Circle, diameter d	$\dfrac{\pi}{16}d^3 S_v$	$\dfrac{M_t}{GJ} = \dfrac{32}{\pi d^4}\dfrac{M_t}{R}$	$2\dfrac{S_{v\max}}{G}\dfrac{1}{d}$	$\dfrac{1}{4}\dfrac{S_{v\max}^2}{G}V$ (Note 1)
Hollow circle, D, d	$\dfrac{\pi}{16}\dfrac{D^4-d^4}{D}S_v$	$\dfrac{32}{\pi(D^4-d^4)}\dfrac{M_t}{G}$	$2\dfrac{S_{v\max}}{G}\dfrac{1}{D}$	$\dfrac{1}{4}\dfrac{S_{v\max}^2}{G}\dfrac{D^2+d^2}{D^2}V$ (Note 2)
Ellipse, axes h, b ($h>b$)	$\dfrac{\pi}{16}b^2 h S_v$	$\dfrac{16}{\pi}\dfrac{b^2+h^2}{b^3 h^3}\dfrac{M_t}{G}$	$\dfrac{S_{v\max}}{G}\dfrac{b^2+h^2}{bh^2}$	$\dfrac{1}{8}\dfrac{S_{v\max}^2}{G}\dfrac{b^2+h^2}{h^2}V$ (Note 3)
Rectangle h, b ($h>b$)	$\dfrac{2}{9}b^2 h S_v$	$3.6\dfrac{b^2+h^2}{b^3 h^3}\dfrac{M_t}{G}$ *	$0.8\dfrac{S_{v\max}}{G}\dfrac{b^2+h^2}{bh^2}$ *	$\dfrac{4}{45}\dfrac{S_{v\max}^2}{G}\dfrac{b^2+h^2}{h^2}V$ (Note 4)
Square, side h	$\dfrac{2}{9}h^3 S_v$	$7.2\dfrac{1}{h^4}\dfrac{M_t}{G}$	$1.6\dfrac{S_{v\max}}{G}\dfrac{1}{h}$	$\dfrac{8}{45}\dfrac{S_{v\max}^2}{G}V$ (Note 5)
Equilateral triangle, side b	$\dfrac{b^3}{20}S_v$	$4.62\dfrac{1}{b^4}\dfrac{M_t}{G}$	$2.31\dfrac{S_{v\max}}{G}\dfrac{1}{b}$	
Hexagon, side b	$\dfrac{b^3}{1.09}S_v$	$0.967\dfrac{1}{b^4}\dfrac{M_t}{G}$	$0.9\dfrac{S_{v\max}}{G}\dfrac{1}{b}$	

*When

$h/b =$	1	2	4	8
Coefficient 3.6 becomes =	3.56	3.50	3.35	3.21
Coefficient 0.8 becomes =	0.79	0.78	0.74	0.71

Notes: (1) $S_{v\max}$ at circumference. (2) $S_{v\max}$ at outer circumference. (3) $S_{v\max}$ at A; $S_{v_B} = 16M_t/\pi bh^2$.
(4) $S_{v\max}$ at middle of side h; in middle of b, $S_v = 9M_t/2bh^2$. (5) $S_{v\max}$ at middle of side.
Source: L. S. Marks (ed.), *Mechanical Engineer's Handbook*, 5th ed., p. 452, McGraw-Hill Book Company, New York, 1951.

Properties of Geometric Sections

TABLE 2.3 Properties of Geometric Sections: Tension, Compression, and Bending Structures

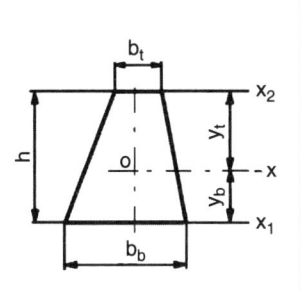

11. TRAPEZOID

$$A = \frac{1}{2}(b_t + b_b)h, \quad y_b = \frac{b_b + 2b_t}{3(b_b + b_t)}h, \quad y_t = \frac{2b_b + b_t}{3(b_b + b_t)}h,$$

$$I_x = \frac{h^3(b_b^2 + 4b_bb_t + b_t^2)}{36(b_b + b_t)}, \quad I_{x_1} = \frac{h^3(b_b + 3b_t)}{12},$$

$$I_{x_2} = \frac{h^3(3b_b + b_t)}{12}, \quad S_{x_1} = \frac{I_x}{y_b} \text{(bottom)}, \quad S_{x_1} = \frac{I_x}{y_t} \text{(top)},$$

$$r_x = \frac{h\sqrt{2(b_b^2 + 4b_bb_t + b_t^2)}}{6(b_b + b_t)}.$$

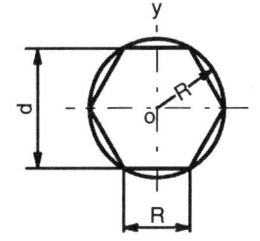

12. REGULAR HEXAGON

$$A = 2.598R^2 = 0.866d^2, \quad I_x = I_y = 0.541R^4 = 0.06d^4,$$

$$S_x = 0.625R^3, \quad S_y = 0.541R^3,$$

$$r_x = r_y = 0.456R = 0.263d.$$

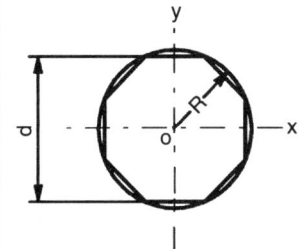

13. REGULAR OCTAGON

$$A = 0.828d^2, \quad I_x = I_y = 0.638R^4 = 0.0547d^4,$$

$$S_x = S_y = 0.690R^3 = 0.1095d^3, \quad r_x = r_y = 0.257d.$$

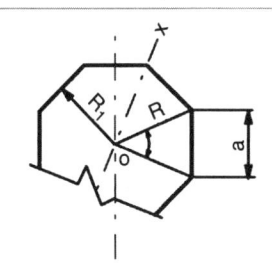

14. REGULAR POLYGON with n sides

$$A = \frac{1}{4}na^2 \cot\frac{\alpha}{2}, \quad R = \frac{a}{2\sin\frac{\alpha}{2}}, \quad R_1 = \frac{a}{2\tan\frac{\alpha}{2}}, \quad \alpha = \frac{360°}{n},$$

$$I_x = I_{x_1} = \frac{naR_1}{96}(12R_1^2 + a^2) = \frac{A}{48}(12R_1^2 + a^2) = \frac{A}{24}(6R^2 + a^2),$$

$$a = 2\sqrt{R^2 - R_1^2}.$$

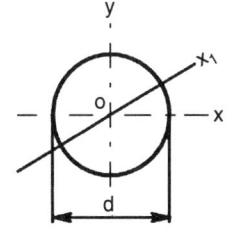

15. CIRCLE

$$A = \frac{\pi d^2}{4} \approx 0.785d^2, \quad I_x = I_y = I_{x_1} = \frac{\pi d^4}{64} \approx 0.05d^4,$$

$$S_x = S_y = S_{x_1} = \frac{\pi d^3}{32} \approx 0.1d^3,$$

$$r_x = r_y = \frac{d}{4}, \quad Z = \frac{d^3}{6}.$$

Shaft Twist and Torque Formulas

Shaft angle-of-twist section formulas

θ = twist, rad
T = torque, in·lb (N·m)
L = length of shaft, in (mm)
N = modulus of rigidity, psi (MPa)
$D, d_o, d_i, d_m, d_s, s, a, b$ = shaft sectional dimensions, in (mm)

Shaft section	Angle of twist θ	Shaft section	Location of max. shear	Torque formulas T
Circle, D	$\dfrac{32TL}{\pi D^4 N}$	Circle, D	Outer fiber	$\dfrac{\pi D^3 f}{16}$
Hollow circle, d_i, d_o	$\dfrac{32TL}{\pi(d_o^4 - d_i^4)N}$	Hollow circle, d_i, d_o	Outer fiber	$\dfrac{\dfrac{\pi}{16}(d_o^4 - d_i^4)}{d_o} f$
Ellipse, d_s, d_m	$\dfrac{16(d_m^2 + d_s^2)TL}{\pi d_m^3 d_s^3 N}$	Ellipse, d_s, d_m	Ends of minor axis	$\dfrac{\pi d_m d_m^2 f}{16}$
Square, s	$\dfrac{7.11 TL}{s^4 N}$	Square, s	Middle of sides	$0.208 S^3 f$
Rectangle, a, b	$\dfrac{3.33(a^2 + b^2)TL}{a^3 b^3 N}$	Rectangle, A, B	Midpoint of major sides	$\dfrac{A^2 B^2 f}{3A + 1.8B}$

FIGURE 2.2 Shaft twist and torque formulas.

Properties of Geometric Sections 37

TABLE 2.4 Properties of Geometric Sections: Tension, Compression, and Bending Structures

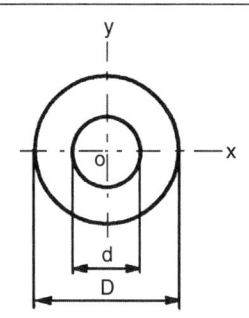

16. HOLLOW CIRCLE

$$A = \frac{\pi D^2}{4}(1-\xi^2), \quad \xi = \frac{d}{D}, \quad I_x = I_y = \frac{\pi D^4}{64}(1-\xi^4),$$

$$S_x = S_y = \frac{\pi D^3}{32}(1-\xi^4), \quad r_x = r_y = \frac{D}{4}\sqrt{1-\xi^2},$$

$$Z = \frac{D^3 - d^3}{6}.$$

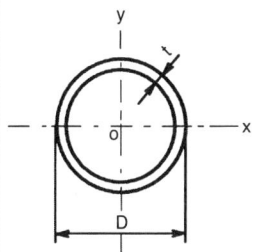

17. THIN RING ($t \ll D$)

$$A = \pi D t, \quad I_x = \frac{\pi D^3 t}{8} \approx 0.3926 D^3 t,$$

$$S_x = \frac{\pi D^2 t}{4} \approx 0.7853 D^2 t, \quad r_x = 0.353 D.$$

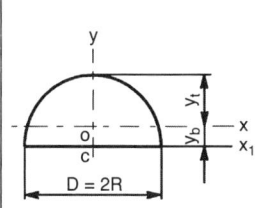

18. Half of a CIRCLE

$$A = \frac{\pi D^2}{8} \approx 0.392 D^2, \quad y_b = 0.2122 D, \quad y_t = 0.2878 D,$$

$$I_x = 0.00686 D^4, \quad I_y = I_{x_1} = \frac{\pi D^4}{128} \approx 0.025 D^4,$$

$$S_{x_b} = 0.2587 \left(\frac{D}{2}\right)^3 \text{ for bottom}, \quad S_{x_t} = 0.1908 \left(\frac{D}{2}\right)^3 \text{ for top}.$$

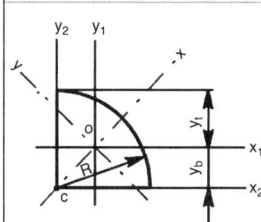

19. Quarter of a CIRCLE

$$A = \frac{\pi R^2}{4} \approx 0.785 R^2, \quad y_b = \frac{4R}{3\pi} \approx 0.424 R, \quad y_t \approx 0.576 R,$$

$$I_x = 0.07135 R^4, \quad I_y = 0.03843 R^4,$$

$$I_{x_1} = I_{y_1} = 0.05489 R^4, \quad I_{x_2} = I_{y_2} = \frac{\pi R^4}{16} \approx 0.19635 R^4.$$

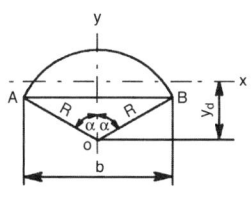

20. Segment of a CIRCLE

$$\hat{\alpha} = \frac{\pi \alpha^0}{180^0}, \quad \varphi = 2\hat{\alpha} - \sin 2\alpha, \quad k = \frac{4 \sin^3 \alpha}{3\varphi}, \quad b = 2R \sin \alpha, \quad s = 2R\hat{\alpha},$$

$$A = \frac{R^2 \varphi}{2}, \quad y_d = kR, \quad I_x = \frac{\varphi R^4}{8}(1 + 3k \cos \alpha), \quad I_y = \frac{\varphi R^4}{8}(1 - k \cos \alpha),$$

($\hat{\alpha}$ in radians, α in degrees).

NOTES

Beam Loading Formulas

Beam Not Loaded in Plane of Symmetry; Flexural Center

The formulas for stress and deflection given in Chap. 3 are valid if the beam is loaded in a plane of symmetry; they are also valid if the applied loads are parallel to either principal central axis of the beam section; but unless the loads also pass through the *elastic axis*, the beam will be subjected to torsion as well as bending.

For the general case of a beam of any section, loaded by a transverse load P in any plane, solution therefore involves the following steps: (1) The load P is resolved into an equal and parallel force P' passing through the flexural center Q of the section, and a twisting couple T equal to the moment of P about Q. (2) P' is resolved at Q into rectangular components P'_u, P'_v, each parallel to a principal central axis of the section. (3) The flexural stresses and deflections due to P'_u, and P'_v are calculated independently by the formulas of Chap. 3 and superposed to find the effect of P'. (4) The stresses due to T are computed independently and superposed on the stresses due to P', giving the stresses due to the actual loading. (Note that T may cause longitudinal fiber stresses as well as shear stresses.) If there are several loads, the effect of each is calculated separately, and these effects are superposed. For a distributed load the same procedure is followed as for a concentrated load.

The above procedure requires the determination of the position of the flexural center Q. For any section having two or more axes of symmetry (rectangle, I beam, etc.) and for any section having a point of symmetry (equilateral triangle, Z bar, etc.), Q is at the centroid. For any section having but one axis of symmetry, Q is on that axis, but in general not at the centroid. For such sections, and for unsymmetrical sections in general, the position of Q must be determined by calculation, by direct experiment, or by the soap-film method.

Table 2.3N gives the position of the flexural center for each of a number of sections.

Properties of Geometric Sections 39

TABLE 2.5 Properties of Geometric Sections: Tension, Compression, and Bending Structures

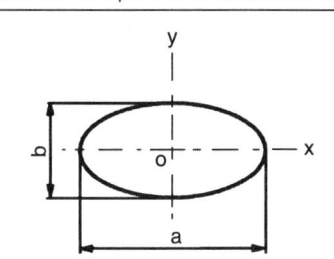

21. ELLIPSE

$$A = \frac{\pi}{4}ab, \quad I_x = \frac{\pi ab^3}{64} = \frac{Ab^2}{16}, \quad I_y = \frac{\pi a^3 b}{64} = \frac{Aa^2}{16},$$

$$S_x = \frac{\pi ab^2}{32} = \frac{Ab}{8}, \quad S_y = \frac{\pi a^2 b}{32} = \frac{Aa}{8},$$

$$r_x = \frac{b}{4}, \quad r_y = \frac{a}{4}.$$

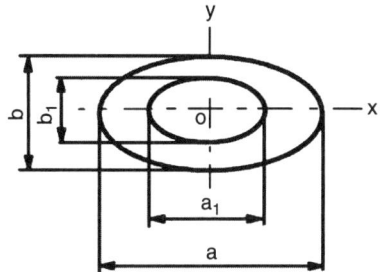

22. HOLLOW ELLIPSE

$$A = \frac{\pi}{4}(ab - a_1 b_1),$$

$$I_x = \frac{\pi}{64}(ab^3 - a_1 b_1^3), \quad I_y = \frac{\pi}{64}(a^3 b - a_1^3 b_1),$$

$$S_x = \frac{\pi}{32b}(ab^3 - a_1 b_1^3), \quad S_y = \frac{\pi}{32a}(a^3 b - a_1^3 b_1).$$

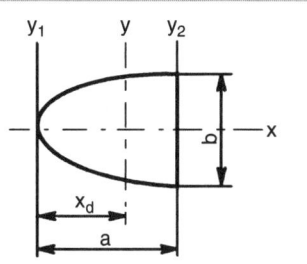

23. Segment of a PARABOLA

$$A = \frac{4ab}{3}, \quad x_d = \frac{3a}{5}, \quad I_x = \frac{4ab^3}{15} = \frac{ab^2}{5},$$

$$I_y = \frac{16a^3 b}{175} = \frac{12Aa^2}{175}, \quad I_{y_1} = \frac{4a^3 b}{7} = \frac{3Aa^2}{7},$$

$$I_{y_2} = \frac{32a^3 b}{105} = \frac{8Aa^2}{35}.$$

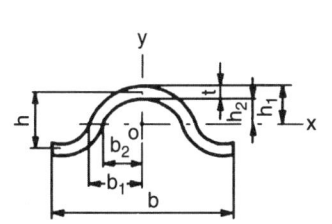

24. STEEL WAVES from parabolic arches

$$A \approx \frac{1}{3}t(2b + 5.2h), \quad b_1 = \frac{1}{4}(b + 2.6t),$$

$$b_2 = \frac{1}{4}(b - 2.6t), \quad h_1 = \frac{1}{2}(h + t),$$

$$h_2 = \frac{1}{2}(h - t), \quad I_x = \frac{64}{105}(b_1 h_1^3 - b_2 h_2^3), \quad S_x \approx \frac{2I_x}{h + t}.$$

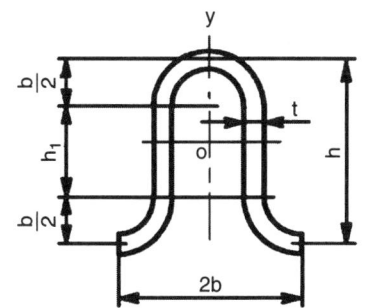

25. STEEL WAVES from circular arches

$$A = (\pi b + 2h)t, \quad h_1 = h - b,$$

$$I_x = \left(\frac{\pi b^3}{8} + b^2 h_1 + \frac{\pi b h_1^2}{4} + \frac{1}{6}h_1^3\right)t,$$

$$S_x = \frac{2I_x}{h + t}.$$

─── N O T E S ───

TABLE 2.3N* Position of Flexural Center Q for Different Sections

Form of Section	Position of Q
1. Any narrow section symmetrical about the x axis. Centroid at $x = 0$, $y = 0$	$e = \dfrac{1+3v}{1+v} \dfrac{\int x'^3 dx}{\int t^3 dx}$ For narrow triangle (with $v = 0.25$), $e = 0.187a$ For any equilateral triangle, $e = 0$ (Refs. 1, 2)
2. Sector of thin circular tube	$e = \dfrac{2R}{(\pi - \theta) + \sin\theta \cos\theta}[(\pi - \theta)\cos\theta + \sin\theta]$ For complete tube split along element ($\theta = 0$) $e = 2R$
3. Semicircular area	$e = \left(\dfrac{8}{15\pi} \dfrac{3+4v}{1+v}\right) R$ (Q is to right of centroid) (Refs. 3, 4) For sector of solid or hollow circular area, see Ref. 1
4. Angle	Leg 1 = rectangle $w_1 h_1$; leg 2 = rectangle $w_2 h_2$ I_1 = moment of inertia of leg 1 about Y_1 (central axis) I_2 = moment of inertia of leg 2 about Y_2 (central axis) $e_x = \dfrac{1}{2} h_2 \left(\dfrac{I_2}{I_1 + I_2}\right)$ $e_y = \dfrac{1}{2} h_1 \left(\dfrac{I_1}{I_1 + I_2}\right)$ (Ref. 5) If w_1 and w_2 are small, $e_x = e_y = 0$ (practically) and Q is at 0
5. Channel	$e = h\left(\dfrac{H_{xy}}{I_x}\right)$ where H_{xy} = product of inertia of the half section (above X) with respect to axes X and Y, and I_x = moment of inertia of whole section with respect to axis X If t is uniform, $e = \dfrac{b^2 h^2 t}{4 I_x}$

---

*From Roark, *Formulas for Stress and Stress and Strain*, 4th ed., McGraw-Hill.
1. A. W. Young, E. M. Elderton, and K. Pearson, "On the Torsion Resulting from Flexure in Prisms with Cross-sections of Uniaxial Symmetry," Drapers' Co. Research Memoirs, Technical Series VII, 1918.
2. W. J. Duncan, The Flexural Center or Center of Shear, *J. Roy. Aeron. Soc.*, vol. 57, September 1953.
3. S. Timoshenko, *Theory of Elasticity*, Engineering Societies Monograph, McGraw-Hill Book Company, 1934.
4. T. Leko, On the Bending Problem of Prismatical Beam by Terminal Transverse Load, *Am. Soc. Mech. Eng., J. Appl. Mech.*, vol. 32, no. 1, March 1965.
5. W. L. S. Schwalbe, The Center of Torsion for Angle and Channel Sections, *Trans. Am. Soc. Mech. Eng.*, vol. 54, no. 11, p. 125, 1932.

Properties of Geometric Sections

TABLE 2.6 Properties of Geometric Sections: Torsion Structures

Cross Section	Moment of Inertia (I_t)	Elastic Section Modulus (S_t)	Position of τ_{max} ($\tau_{max} = M_t/S_t$)
Circle, diameter d	$I_t = \dfrac{\pi d^4}{32} = I_p$	$S_t = \dfrac{\pi d^3}{16}$	At all points of the perimeter
Hollow circle, d_1, d_2	$I_t = \dfrac{\pi}{32} \cdot (d_2^4 - d_1^4) = I_p$	$S_t = \dfrac{\pi}{16} \cdot \dfrac{d_2^4 - d_1^4}{d_2}$	At all points of the outside perimeter
Hexagon inscribed in circle d, R	$I_t = 0.1154 d^4$	$S_t = 0.1888 d^3$	In the middle of the sides
Octagon inscribed in circle d	$I_t = 0.1075 d^4$	$S_t = 0.1850 d^3$	In the middle of the sides
Square, side a	$I_t = 0.1404 a^4$	$S_t = 0.208 a^3$	In the middle of the sides
Trapezoid h, b_1, b_2	$I_t = \dfrac{h(b_1^4 - b_2^4)}{12(b_1 - b_2)} - 0.21 b_2^4$	$S_t = \dfrac{I_t}{b_1}$	In the middle of the long side

W shape	Angle	Channel	Structural tee
$I_t = \eta \cdot \sum\limits_{i=1}^{i=n} \dfrac{h_i b_i^3}{3}$		$S_t = \dfrac{I_t}{b_{max}}$	
I	L	[	⊥
$n = 3$, $\eta = 1.2$	$n = 2$, $\eta = 1.0$	$n = 3$, $\eta = 1.12$	$n = 2$, $\eta = 1.15$

Torsion in Solid and Hollow Shafts

A 30-ft (9.1-m) long solid shaft weighing 150 lb/ft (223.2 kg/m) is fitted with a pulley and a gear. The gear delivers 100 hp (74.6 kW) to the shaft while driving the shaft at 500 r/min. Determine the required diameter of the shaft if the allowable stress is 10,000 lb/in² (68,947.6 kPa).

Calculation Procedure

1. *Compute the pulley and gear concentrated loads.* Using the method of the previous calculation procedure, we get $T = 63,000 \text{hp}/R = 63,000(100)/500 = 12,600$ lb·in (1423·6 N·m). Assuming that the maximum tension of the tight side of the belt is twice the tension of the slack side, we see the maximum belt load is $R_p = 3T/r = 3(12,600)/24 = 1575$ lb (7005.9 N). Hence, the total pulley concentrated load = belt load + pulley weight = 1575 + 750 = 2325 lb (10,342.1 N).

   The gear concentrated load is found from $F_g = T/r$, where the torque is the same as computed for the pulley, or $F_g = 12,600/9 = 1400$ lb (6227.5 N). Hence, the total gear concentrated load is 1400 + 75 = 1475 lb (6561.1 N).

   Draw a sketch of the shaft, showing the two concentrated loads in position.

2. *Compute the end reactions of the shaft.* Take moments about R_R to determine L_R, using the method of the previous calculation procedures. Thus, $L_R(30) - 2325(25) - 1475(8) - 150(30)(15) = 0$.

3. *Analyze the hollow shaft.* The usual practice is to size hollow shafts such that the ratio q of the inside diameter d_i into the outside diameter d_o in is 1:2 to 1:3 or some intermediate value. With a q in this range the shaft will have sufficient thickness to prevent failure in service.

   Assume $q = d_i/d_o = 1/2$. Then with $d_i = 1.0$ in (2.5 cm), $d_o = d_i/q$, or $d_o = 1.0/0.5 = 2.0$ in (5.1 cm). With $q = 1/3$, $d_o = 1.0/0.33 = 3.0$ in (7.6 cm).

4. *Compute the stress in each hollow shaft.* For the hollow shaft $s = 5.1\, T/d_o^3,\, (1 - q^4)$, where the symbols are as defined above. Thus, for the 2-in (5.1-cm) outside-diameter shaft, $s = 5.1\,(8750)/[8(1 - 0.0625)] = 5950$ lb/in² (41,023.8 kPa).

   By inspection, the stress in the 3-in (7.6-cm) outside-diameter shaft will be lower because the torque is constant. Thus, $s = 5.1(8750)/[27(1 - 0.0123)] = 1672$ lb/in² (11,528.0 kPa).

5. *Choose the outside diameter of the hollow shaft.* Use a trial-and-error procedure to choose the hollow shaft's outside diameter. Since the stress in the 2-in (5.1-cm) outside-diameter shaft, 5950 lb/in² (41,023.8 kPa), is less than one-half the allowable

stress of 12,500 lb/in² (86,187.5 kPa), select a smaller outside diameter and compute the stress while holding the inside diameter constant.

Thus, with a 1.5-in (3.8-cm) shaft and the same inside diameter, $s = 5.1(8750)/[3.38(1 - 0.197)] = 16{,}430$ lb/in² (113,284.9 kPa). This exceeds the allowable stress.

Try a larger outside diameter, 1.75 in (4.4 cm), to find the effect on the stress. Or $s = 5.1(8750)/[(5.35(1 - 0.107)] = 9350$ lb/in² (64,468.3 kPa). This is lower than the allowable stress.

Since a 1.5-in (3.8-cm) shaft has a 16,430 lb/in² (113,284.9 kPa) stress and a 1.75-in (4.4-cm) shaft has a 9350-lb/in² (64,468.3 kPa) stress, a shaft of intermediate size will have a stress approaching 12,500 lb/in² (86,187.5 kPa). Trying 1.625 in (4.1 cm) gives $s = 5.1(8750)/[4.4(1 - 0.143)] = 11{,}820$ lb/in² (81,489.9 kPa). This is within 680 lb/in² (4688.6 kPa) of the allowable stress and is close enough for usual design calculations.

PART II

Statics

CHAPTER 3
Beams: Diagrams and Formulas for Various Loading Conditions

CHAPTER 4
Frames: Diagrams and Formulas for Various Static Loading Conditions

CHAPTER 5
Arches: Diagrams and Formulas for Various Loading Conditions

CHAPTER 6
Trusses: Method of Joints and Method of Section Analysis

CHAPTER 7
Plates: Bending Moments for Various Support and Loading Conditions

CHAPTER 3
Beams: Diagrams and Formulas for Various Loading Conditions

NOTES

The formulas provided in Tables 3.1 to 3.10—for determination of support reactions (R), bending moments (M), and shears (V)—are to be used for elastic beams with constant or variable cross sections.

The formulas for determination of deflection and angles of deflection can only be used for elastic beams with constant cross sections.

Beams: Diagrams and Formulas for Various Loading Conditions

TABLE 3.1 Simple Beams

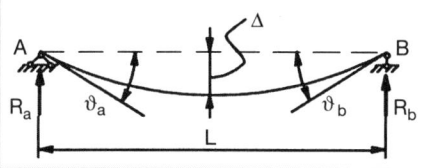

Notes:
$V_1 = R_a$, $V_2 = R_b$
θ_a and θ_b in radians

Loadings	Support Reactions	Bending Moment	Deflection	Angle of Deflection
	$R_a = \dfrac{P}{2}$ $R_b = \dfrac{P}{2}$	$M_{max} = \dfrac{PL}{4}$ at point of load	$\Delta_{max} = \dfrac{PL^3}{48EI}$ at point of load	$\theta_a = \theta_b = \dfrac{PL^2}{16EI}$
	$R_a = P\dfrac{b}{L}$ $R_b = P\dfrac{a}{L}$	$M_{max} = P\dfrac{ab}{L}$ at point of load	$\Delta_a = \dfrac{Pa^2b^2}{3EI \cdot L}$ at point of load	$\theta_a = \dfrac{PL^2}{6EI}(\xi_1 - \xi_1^3)$ $\theta_b = \dfrac{PL^2}{6EI}(\xi - \xi^3)$ $\xi = \dfrac{a}{L}, \xi_1 = \dfrac{b}{L}$
	$R_a = R_b = P$	$M_{max} = Pa$ between loads	$\Delta_{max} = \dfrac{Pa(3L^2 - 4a^2)}{24EI}$ at center	$\theta_a = \theta_b = \dfrac{Pa(L-a)}{2EI}$
	$R_a = \dfrac{3P}{2}$ $R_b = \dfrac{3P}{2}$	$M_{max} = \dfrac{PL}{2}$ at center	$\Delta_{max} = \dfrac{PL^3}{20.22EI}$ at center	$\theta_a = \theta_b = 3.75\dfrac{PL^2}{24EI}$

─ NOTES ─

Example for Table 3.2. Computation of beam

Given. Simple beam $W14 \times 145$, $L = 10$ m

Moment of inertia $I = 1710$ in$^4 \times 2.54^4 = 71{,}175.6$ cm^4

Modulus of elasticity $E = 29{,}000$ kip/in$^2 = \dfrac{29{,}000 \times 4.48222}{2.54^2}$

$= 20{,}147.6$ kN/cm^2

Uniform distribution load $w = 5$ kN/m $= 0.05$ kN/cm

Required. Compute $V = R$, M_{max}, Δ_{max}, $\theta = \theta_a = \theta_b$.

Solution. $V = R = \dfrac{wL}{2} = \dfrac{5 \times 10}{2} = 25$ kN

$M_{max} = \dfrac{wL^2}{8} = \dfrac{5 \times 10^2}{8} = 62.5$ kN·m

$\Delta_{max} = \dfrac{5}{384} \cdot \dfrac{wL^4}{EI} = \dfrac{5}{384} \cdot \dfrac{0.05 \times (1000)^4}{20{,}147.6 \times 71{,}175.6} = 0.45$ cm $= 4.5$ mm

$\theta = \dfrac{wL^3}{24EI} = \dfrac{0.05 \times (1000)^3}{24 \times 20{,}147.6 \times 71{,}175.6} = 1.45 \times 10^{-3}$ rad

Beams: Diagrams and Formulas for Various Loading Conditions

TABLE 3.2 Simple Beams

Loadings	Support Reactions	Bending Moment	Deflection			Angle of Deflection
		$n = 4$	5	6		
n equal loads	$R_a = \dfrac{Pn}{2}$	$M_{max} = \dfrac{PL}{2}$	$\dfrac{PL}{1.538}$	$\dfrac{PL}{1.333}$		$\theta_a = \dfrac{PL^2}{48EI} \cdot \dfrac{2n^2+1}{n}$
	$R_b = \dfrac{Pn}{2}$	$\Delta_{max} = \dfrac{PL^3}{19.04EI}$	$\dfrac{PL^3}{15.1EI}$	$\dfrac{PL^3}{12.65EI}$		$\theta_b = \dfrac{PL^2}{48EI} \cdot \dfrac{2n^2+1}{n}$
(uniform load w)	$R_a = \dfrac{wL}{2}$	$M_{max} = \dfrac{wL^2}{8}$ at center	$\Delta_{max} = \dfrac{5}{384} \cdot \dfrac{wL^4}{EI}$ at center			$\theta_a = \theta_b = \dfrac{wL^3}{24EI}$
	$R_b = \dfrac{wL}{2}$	$M_x = \dfrac{wx}{2}(L-x)$	$\Delta_x = \dfrac{wx(L^3 - 2Lx^2 + x^3)}{24EI}$			
$a = \xi L$, b	$R_a = \dfrac{wa}{2}(2-\xi)$	$M_{max} = \dfrac{wa^2}{8}(2-\xi)^2$	$\Delta_a = \dfrac{wa^3 b}{24EI}(4-3\xi)$			$\theta_a = \dfrac{wa^2 L}{6EI}\left(1 - \dfrac{1}{2}\xi\right)^2$
	$R_b = \dfrac{wa}{2}\cdot \xi$	at $x = \dfrac{a}{2}(2-\xi)$	at $x = a$			$\theta_b = \dfrac{wa^2 L}{12EI}\left(1 - \dfrac{1}{2}\xi^2\right)$
	$\xi = \dfrac{a}{L}$					
a, b	$R_a = \dfrac{wcb}{L}$	$M_{max} = \dfrac{wabc}{L}\left(1 - \dfrac{c}{2L}\right)$	$\Delta_a = \left[a\left(2aL - 2a^2 - \dfrac{c^2}{4}\right)\right.$			$\theta_a = \dfrac{R_a}{24EI}\cdot f_1$
	$R_b = \dfrac{wca}{L}$	at $x = a + \dfrac{c(b-a)}{2L}$	$\left. + \dfrac{c^3 L}{64b}\right] \times \dfrac{R_a}{6EI}$			$\theta_b = \dfrac{R_b}{24EI}\cdot f_1$
			at $x = a$			$f_1 = 4a(L+b) - c^2$

Safe Loads for Beams of Various Types

Table 3.1N gives 32 formulas for computing the approximate safe loads on steel beams of various cross sections for an allowable stress of 16,000 lb/in² (110.3 MPa). Use these formulas for quick estimation of the safe load for any steel beam you are using in a design.

Table 3.2N gives coefficients for correcting values in Table 3.1N for various methods of support and loading. When combined with Table 3.1N, the two sets of formulas provide useful time-saving means of making quick safe-load computations in both the office and the field.

TABLE 3.1N Approximate Safe Loads in Pounds (kgf) on Steel Beams* (Percoyd Iron Works)
(Beam supported at both ends; allowable fiber stress for steel, 16,000 lb/in² (1.127 kgf/cm²) (basis of table) for iron, reduce values given in table by one-eighth)

	Greatest Safe Load, lb[†]		Deflection, in[†]	
Shape of Section	Load in Middle	Load Distributed	Load in Middle	Load Distributed
Solid rectangle	$\dfrac{890AD}{L}$	$\dfrac{1780AD}{L}$	$\dfrac{wL^3}{32AD^2}$	$\dfrac{wL^3}{52AD^2}$
Hollow rectangle	$\dfrac{890(AD-ad)}{L}$	$\dfrac{1780(AD-ad)}{L}$	$\dfrac{wL^3}{32(AD^2-ad^2)}$	$\dfrac{wL^3}{52(AD^2-ad^2)}$
Solid cylinder	$\dfrac{667AD}{L}$	$\dfrac{1333AD}{L}$	$\dfrac{wL^3}{24AD^2}$	$\dfrac{wL^3}{38AD^2}$
Hollow cylinder	$\dfrac{667(AD-ad)}{L}$	$\dfrac{1333(AD-ad)}{L}$	$\dfrac{wL^3}{24(AD^2-ad^2)}$	$\dfrac{wL^3}{38(AD^2-ad^2)}$
Even-legged angle or tee	$\dfrac{885AD}{L}$	$\dfrac{1770AD}{L}$	$\dfrac{wL^3}{32AD^2}$	$\dfrac{wL^3}{52AD^2}$
Channel or Z bar	$\dfrac{1525AD}{L}$	$\dfrac{3050AD}{L}$	$\dfrac{wL^3}{53AD^2}$	$\dfrac{wL^3}{85AD^2}$
Deck beam	$\dfrac{1380AD}{L}$	$\dfrac{2760AD}{L}$	$\dfrac{wL^3}{50AD^2}$	$\dfrac{wL^3}{80AD^2}$
I beam	$\dfrac{1795AD}{L}$	$\dfrac{3390AD}{L}$	$\dfrac{wL^3}{58AD^2}$	$\dfrac{wL^3}{93AD^2}$

*L = distance between supports, ft (m); A = sectional area of beam, in² (cm²); D = depth of beam, in (cm); a = interior area, in² (cm²); d = interior depth, in (cm); w = total working load, net tons (kgf).

[†]See Table 3.2N for coefficients for correcting values for various methods of support and loading.

TABLE 3.3 Simple Beams

Loadings	Support Reactions	Bending Moment	Deflection	Angle of Deflection
(triangular load)	$R_a = \dfrac{wL}{6}$ $R_b = \dfrac{wL}{3}$	$M_{max} = \dfrac{wL^2}{9\sqrt{3}} = 0.064wL^2$ when $x = 0.577L$	$\Delta_{max} = 0.00652\dfrac{wL^4}{EI}$ when $x = 0.519L$	$\theta_a = \dfrac{7}{360}\cdot\dfrac{wL^3}{EI}$ $\theta_b = \dfrac{8}{360}\cdot\dfrac{wL^3}{EI}$
(symmetric triangular)	$R_a = R_b = \dfrac{wL}{4}$	$M_{max} = \dfrac{wL^2}{12}$ at center	$\Delta_{max} = \dfrac{wL^2}{120EI}$ at center	$\theta_a = \theta_b = \dfrac{5wL^3}{192EI}$
(trapezoidal load)	$R_a = \dfrac{w(L-a)}{2}$ $R_b = \dfrac{w(L-a)}{2}$	$M_{max} = \dfrac{wL^2}{8} - \dfrac{wa^2}{6}$ at center	$\Delta_{max} = \dfrac{5}{384}\cdot\dfrac{wL^4}{EI}\cdot f_2$ $f_2 = 1 - \dfrac{8}{5}\xi^2 + \dfrac{16}{25}\xi^4$ at center	$\theta_a = \theta_b = \dfrac{wL^3}{24EI}\cdot f_3$ $f_3 = 1 - 2\xi^2 + \xi^3$

Loadings	Support Reactions	Bending Moment						Angle of Deflection
(asymmetric trapezoid)	$R_a = \dfrac{2w_a + w_b}{6}L$ $R_b = \dfrac{w_a + 2w_b}{6}L$	$\dfrac{w_a}{w_b} =$	0.2	0.4	0.6	0.8	1.0	$\theta_a = \dfrac{L^3(8w_a + 7w_b)}{360EI}$ $\theta_b = \dfrac{L^3(7w_a + 8w_b)}{360EI}$
		$M_{max} =$	$\dfrac{w_bL^2}{13.09}$	$\dfrac{w_bL^2}{11.30}$	$\dfrac{w_bL^2}{9.93}$	$\dfrac{w_bL^2}{8.87}$	$\dfrac{w_bL^2}{8.00}$	
		$\dfrac{x}{L} =$	0.555	0.536	0.520	0.508	0.500	
		$\Delta_{max} = (w_a + w_b)L^4$, when $x = 0.500L$ to $x = 0.519L$						

NOTES

TABLE 3.2N Coefficients for Correcting Values in Table 3.1N for Various Methods of Support and Loading*

Conditions of Loading	Max. Relative Safe Load	Max. Relative Deflection Under Max. Relative Safe Load
Beam supported at ends		
Load uniformly distributed over span	1.0	1.0
Load concentrated at center of span	1/2	0.80
Two equal loads symmetrically concentrated	$l/4c$	
Load increasing uniformly to one end	0.974	0.976
Load increasing uniformly to center	3/4	0.96
Load decreasing uniformly to center	3/2	1.08
Beam fixed at one end, cantilever		
Load uniformly distributed over span	1/4	2.40
Load concentrated at end	1/8	3.20
Load increasing uniformly to fixed end	3/8	1.92
Beam continuous over two supports equidistant from ends		
Load uniformly distributed over span 1. If distance $a > 0.2071l$ 2. If distance $a < 0.2071l$ 3. If distance $a = 0.2071l$	$l^2/4a^2$ $l/(l - 4a)$ 5.83	
Two equal loads concentrated at ends	$l/4a$	

*l = length of beam; c = distance from support to nearest concentrated load; a = distance from support to end of beam.

Beams: Diagrams and Formulas for Various Loading Conditions

TABLE 3.4 Simple Beams and Beams Overhanging One Support

Loadings	Support Reactions	Bending Moment	Deflection	Angle of Deflection
(Moment M_a at end)	$R_a = \dfrac{M_a}{L}$ $R_b = -R_a$	$M_{max} = M_a$ when $x = 0$	$\Delta_{max} = \dfrac{M_a L^2}{15.59 EI}$ when $x = 0.423L$ $\Delta = \dfrac{M_a L^2}{16 EI}$ when $x = 0.5L$	$\theta_a = \dfrac{M_a L}{3EI}$ $\theta_b = \dfrac{M_a L}{6EI}$
(Moment M_0 at interior point)	$R_a = \dfrac{M_0}{L}$ $R_b = \dfrac{M_0}{L}$	$M_1 = -M_0 \dfrac{a}{L}$ $M_2 = M_0 \dfrac{b}{L}$	$\Delta = \dfrac{M_0 ab}{3EI}\left(\dfrac{a-b}{L}\right)$ when $x = a$	$\theta_a = -\dfrac{M_0 L}{6EI} f_4$ $\theta_b = \dfrac{M_0 L}{6EI} f_5$ $f_4 = 1 - 3\left(\dfrac{b}{L}\right)^2$ $f_5 = 1 - 3\left(\dfrac{a}{L}\right)^2$
(Load P at overhang end)	$R_a = -P\dfrac{a}{L}$ $R_b = P\dfrac{a+L}{L}$	$M_b = -Pa$	For overhang: $\Delta = \dfrac{Pa^2}{3EI}(L+a)$ Between supports: $\Delta_{max} = -0.0642 \dfrac{PaL^2}{EI}$, $x = 0.577L$	For overhang: $\theta = \dfrac{P(2aL + 3a^2)}{6EI}$ $\theta_a = -\dfrac{PaL}{6EI}$ $\theta_b = -\dfrac{PaL}{3EI}$
(UDL w on overhang)	$R_a = -\dfrac{wa^2}{2L}$ $R_b = w\left(a + \dfrac{a^2}{2L}\right)$	$M_b = -\dfrac{wa^2}{2}$	For overhang: $\Delta = \dfrac{wa^3}{24EI}(4L + 3a)$ Between supports: $\Delta_{max} = -0.0321 \dfrac{wa^2 L^2}{EI}$, $x = 0.577L$	For overhang: $\theta = \dfrac{wa^2(a+L)}{6EI}$ $\theta_a = -\dfrac{wa^2 L}{12EI}$ $\theta_b = -\dfrac{wa^2 L}{6EI}$

NOTES

Torsion in Structural Members

Torsion in structural members occurs when forces or moments twist the beam or column. For circular members, Hooke's law gives the shear stress at any given radius r. Table 3.3N shows the polar moment of inertia J and the maximum shear for five different structural sections.

TABLE 3.3N Polar Moment of Inertia and Maximum Torsional Shear

Section	Polar Moment of Inertia J	Maximum Shear* v_{max}
Circle, $2r$	$\frac{1}{2}\pi r^4$	$\frac{2T}{\pi r^3}$ at periphery
Square, a	$0.141a^4$	$R = \frac{2T}{208a^3}$ at midpoint of each side
Rectangle, $a \times b$	$ab^3\left[\frac{1}{3} - 0.21\frac{b}{a}\left(1 - \frac{b^4}{12a^4}\right)\right]$	$\frac{T(3a - 1.8b)}{a^2b^2}$ at midpoint of longer sides
Triangle, a	$0.0214a^4$	$\frac{20T}{a^3}$ at midpoint of each side
Hollow circle, $2r$ $2R$	$\frac{1}{2}\pi(R^4 - r^4)$	$\frac{2TR}{\pi(R^4 - r^4)}$ at outer periphery

*T = twisting moment, or torque.

Beams: Diagrams and Formulas for Various Loading Conditions

TABLE 3.5 Cantilever Beams

Loadings	Reaction (at Fixed End)	Bending Moment (at Fixed End)	Deflection (at Free End)	Angle of Deflection (at Free End)
Point load P at free end	$R = P$	$M_{max} = -PL$	$\Delta_{max} = \dfrac{PL^3}{3EI}$	$\theta = \dfrac{PL^2}{2EI}$
Point load P at distance a from fixed end	$R = P$	$M_{max} = -Pa$	$\Delta_{max} = \dfrac{Pa^2}{6EI}(3L-a)$	$\theta = \dfrac{Pa^2}{2EI}$
Uniform load w	$R = wL$	$M_{max} = -\dfrac{wL^2}{2}$	$\Delta_{max} = \dfrac{wL^4}{8EI}$	$\theta = \dfrac{wL^3}{6EI}$
Triangular load (max w at fixed end)	$R = \dfrac{wL}{2}$	$M_{max} = -\dfrac{wL^2}{6}$	$\Delta_{max} = \dfrac{wL^4}{30EI}$	$\theta = \dfrac{wL^3}{24EI}$

Eccentric Loading

If an eccentric longitudinal load is applied to a bar in the plane of symmetry, it produces a bending moment Pe, where e is distance, in (mm), of the load P from the centroidal axis. The total unit stress is the sum of this moment and the stress due to P applied as an axial load:

$$f = \frac{P}{A} \pm \frac{Pec}{I} = \frac{P}{A}\left(1 \pm \frac{ec}{r^2}\right) \tag{3.1}$$

where A = cross-sectional area, in² (mm²)
c = distance from neutral axis to outermost fiber, in (mm)
I = moment of inertia of cross section about neutral axis, in⁴ (mm⁴)
r = radius of gyration = $\sqrt{I/A}$, in (mm)

If there is to be no tension on the cross section under a compressive load, e should not exceed r^2/c. For a rectangular section with width b and depth d, the eccentricity, therefore, should be less than $b/6$ and $d/6$ (i.e., the load should not be applied outside the middle third). For a circular cross section with diameter D, the eccentricity should not exceed $D/8$.

When the eccentric longitudinal load produces a deflection too large to be neglected in computing the bending stress, account must be taken of the additional bending moment Pd, where d is the deflection, in (mm). This deflection may be closely approximated by

$$d = \frac{4eP/P_c}{\pi(1 - P/P_c)} \tag{3.2}$$

and P_c is the critical buckling load $\pi^2\, EI/L^2$, lb (N).

If the load P does not lie in a plane containing an axis of symmetry, it produces bending about the two principal axes through the centroid, of the section. The stress, lb/in² (MPa), are given by

$$f = \frac{P}{A} + \frac{Pe_x c_x}{I_y} + \frac{Pe_y c_y}{I_x} \tag{3.3}$$

where A = cross-sectional area, in² (mm²)
e_x = eccentricity with respect to principal axis YY, in (mm)
e_y = eccentricity with respect to principal axis XX, in (mm)
c_x = distance from YY to outermost fiber, in (mm)
c_y = distance from XX to outermost fiber, in (mm)
I_x = moment of inertia about XX, in⁴ (mm⁴)
I_y = moment of inertia about YY, in⁴ (mm⁴)

The principal axes are the two perpendicular axes through the centroid for which the moments of inertia are a maximum or a minimum and for which the products of inertia are zero.

Beams: Diagrams and Formulas for Various Loading Conditions

TABLE 3.6 Beams Fixed at One End, Supported at Other

Loadings	Support Reactions	Bending Moments and Deflection
(concentrated load P at distance a from fixed end, b from support)	$R_a = \dfrac{Pb}{2L^3}(3L^2 - b^2)$ $R_b = \dfrac{Pa^2}{2L^3}(b+2L)$	$M_a = -\dfrac{Pab}{2L^2}(L+b)$, at fixed end $M_1 = R_b b$, at point of load $\Delta_1 = \dfrac{Pa^2 b^2 (3a+4b)}{12L^3 EI}$, at point of load
(uniform load w over full span)	$R_a = \dfrac{5}{8}wL$ $R_b = \dfrac{3}{8}wL$	$M_a = -\dfrac{wL^2}{8}$, at fixed end $M_1 = \dfrac{9}{128}wL^2$, at $x = 0.625L$ $\Delta_{max} = \dfrac{wL^4}{185EI}$, at $x = 0.579L$ $\Delta = \dfrac{wL^4}{192EI}$, at $x = \dfrac{L}{2}$
(triangular load, max w at fixed end)	$R_a = \dfrac{2}{5}wL$ $R_b = \dfrac{1}{10}wL$	$M_a = -\dfrac{wL^2}{15}$, at fixed end $M_1 = \dfrac{wL^2}{33.6}$, at $x = 0.553L$ $\Delta_{max} = \dfrac{wL^4}{419EI}$, at $x = 0.553L$ $\Delta = \dfrac{wL^4}{426.6EI}$, at $x = \dfrac{L}{2}$
(moment M_b applied at support)	$R_a = \dfrac{3}{2} \cdot \dfrac{M_b}{L}$ $R_b = -\dfrac{3}{2} \cdot \dfrac{M_b}{L}$	$M_a = -\dfrac{M_b}{2}$, at fixed end $\Delta_{max} = \dfrac{M_b L^2}{27EI}$, at $x = \dfrac{2}{3}L$

Combined Axial and Bending Loads

For short beams, subjected to both transverse and axial loads, the stresses are given by the principle of superposition if the deflection due to bending may be neglected without serious error. That is, the total stress is given with sufficient accuracy at any section by the sum of the axial stress and the bending stresses. The maximum stress, lb/in² (MPa), equals

$$f = \frac{P}{A} + \frac{Mc}{I} \tag{3.4}$$

where P = axial load, lb (N)
A = cross-sectional area, in² (mm²)
M = maximum bending moment, in · lb (N · m)
c = distance from neutral axis to outermost fiber at the section where maximum moment occurs, in (mm)
I = moment of inertia about neutral axis at that section, in⁴ (mm⁴)

When the deflection due to bending is large and the axial load produces bending stresses that cannot be neglected, the maximum stress is given by

$$f = \frac{P}{A} + (M + Pd)\frac{c}{I} \tag{3.5}$$

where d is the deflection of the beam. For axial compression, the moment Pd should be given the same sign as M; and for tension, the opposite sign, but the minimum value of $M + Pd$ is zero. The deflection d for axial compression and bending can be closely approximated by

$$d = \frac{d_0}{1 - P/P_c} \tag{3.6}$$

where d_0 = deflection for the transverse loading alone, in (mm), and P_c = critical buckling load $\pi^2 EI/L^2$, lb (N).

Unsymmetrical Bending

When a beam is subjected to loads that do not lie in a plane containing a principal axis of each cross section, unsymmetrical bending occurs. Assuming that the bending axis of the beam lies in the plane of the loads, to preclude torsion, and that the loads are perpendicular to the bending axis, to preclude axial components, the stress, lb/in² (MPa), at any point in a cross section is

$$f = \frac{M_x y}{I_x} + \frac{M_y x}{I_y} \tag{3.7}$$

Continued on page 62

Beams: Diagrams and Formulas for Various Loading Conditions 61

TABLE 3.7 Beams Fixed at One End, Supported at Other End

Loadings	Support Reactions	Bending Moment (At Fixed End)
	$R_a = -\dfrac{3M_0(L^2 - b^2)}{2L^3}$ $R_b = \dfrac{3M_0(L^2 - b^2)}{2L^3}$	$M_a = \dfrac{M_0}{2}\left[1 - 3\left(\dfrac{b}{L}\right)^2\right]$, when $b < 0.577L$ $M_a = 0$, when $b = 0.577L$ $M_a = -\dfrac{M_0}{2}\left[1 - 3\left(\dfrac{b}{L}\right)^2\right]$, when $b > 0.577L$
	$R_a = -\dfrac{3EI}{L^3}$ $R_b = \dfrac{3EI}{L^3}$	$M_a = \dfrac{3EI}{L^2}$
	$R_a = \dfrac{3EI}{L^3}$ $R_b = -\dfrac{3EI}{L^3}$	$M_a = -\dfrac{3EI}{L^2}$
	$R_a = \dfrac{3EI}{L^2}$ $R_b = -\dfrac{3EI}{L^2}$	$M_a = -\dfrac{3EI}{L}$

─── NOTES ───

where M_x = bending moment about principal axis XX, in · lb (N · m)
M_y = bending moment about principal axis YY, in · lb (N · m)
x = distance from point where stress is to be computed to YY axis, in (mm)
y = distance from point to XX axis, in (mm)
I_x = moment of inertia of cross section about XX, in (mm⁴)
I_y = moment of inertia about YY, in (mm⁴)

If the plane of the loads makes an angle θ with a principal plane, the neutral surface forms an angle α with the other principal plane such that

$$\tan \alpha = \frac{I_x}{I_y} \tan \theta \tag{3.8}$$

Computation of Fixed-End Moments in Prismatic Beams

Curves (Fig. 3.1) can be used to speed computation of fixed-end moments in prismatic beams. Before the curves in Fig. 3.1 can be used, the characteristics of the loading must be computed by using the formulas in Fig. 3.2. These include $\bar{x}L$, the location of the center of gravity of the loading with respect to one of the loads; $G^2 = \Sigma b_n^2 P_n / W$, where $b_n L$ is the distance from each load P_n to the center of gravity of the loading (taken positive to the right); and $S^3 = \Sigma b_n^3 P_n / W$. These values are given in Fig. 3.2 for some common types of loading.

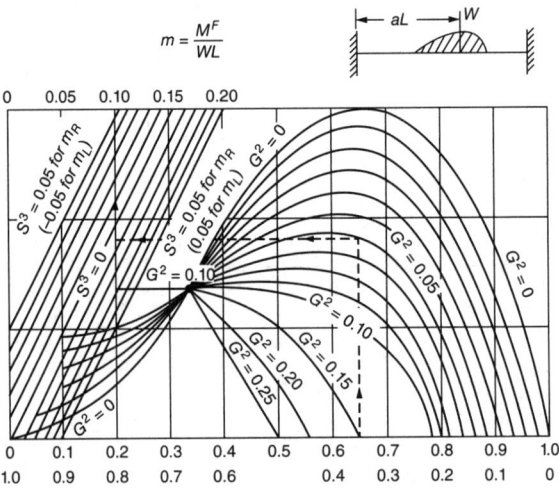

FIGURE 3.1 Chart for fixed-end moments due to any type of loading.

Beams: Diagrams and Formulas for Various Loading Conditions

TABLE 3.8 Beams Fixed at Both Ends

Loadings	Support Reactions	Bending Moments and Deflection
(concentrated load P at distance a from left, b from right)	$R_a = \dfrac{P(3a+b)b^2}{L^3}$ $R_b = \dfrac{P(a+3b)a^2}{L^3}$	$M_a = -\dfrac{Pab^2}{L^2}, \quad M_b = -\dfrac{Pa^2b}{L^2}$ $M_1 = \dfrac{2Pa^2b^2}{L^3}$, at point of load $\Delta_1 = \dfrac{Pa^3b^3}{3L^3EI}$, at point of load
(uniform load w over full span L)	$R_a = R_b = \dfrac{wL}{2}$	$M_a = M_b = -\dfrac{wL^2}{12}$ $M_1 = \dfrac{wL^2}{24}$, at center $\Delta_{max} = \dfrac{wL^4}{384EI}$, at center
(triangular load w over L)	$R_a = \dfrac{7}{20}wL$ $R_b = \dfrac{3}{20}wL$	$M_a = -\dfrac{wL^2}{20}, \quad M_b = -\dfrac{wL^2}{30}$ $M_1 = \dfrac{wL^2}{46.6}$, at $x = 0.452L$ $\Delta_{max} = \dfrac{wL^4}{764EI}$, at $x = 0.475L$ $\Delta = \dfrac{wL^4}{768EI}$, at $x = \dfrac{L}{2}$
(partial uniform load w over length $a = \xi L$)	$R_a = \dfrac{wa(L-0.5a)}{L} - \dfrac{M_a - M_b}{L}$ $R_b = \dfrac{wa^2}{2L} + \dfrac{M_a - M_b}{L}$	$M_a = -\dfrac{wa^2}{6}(3 - 4\xi + 1.5\xi^2)$ $M_b = -\dfrac{wa^2}{3}(\xi - 0.75\xi^2)$ $\xi = \dfrac{a}{L}$

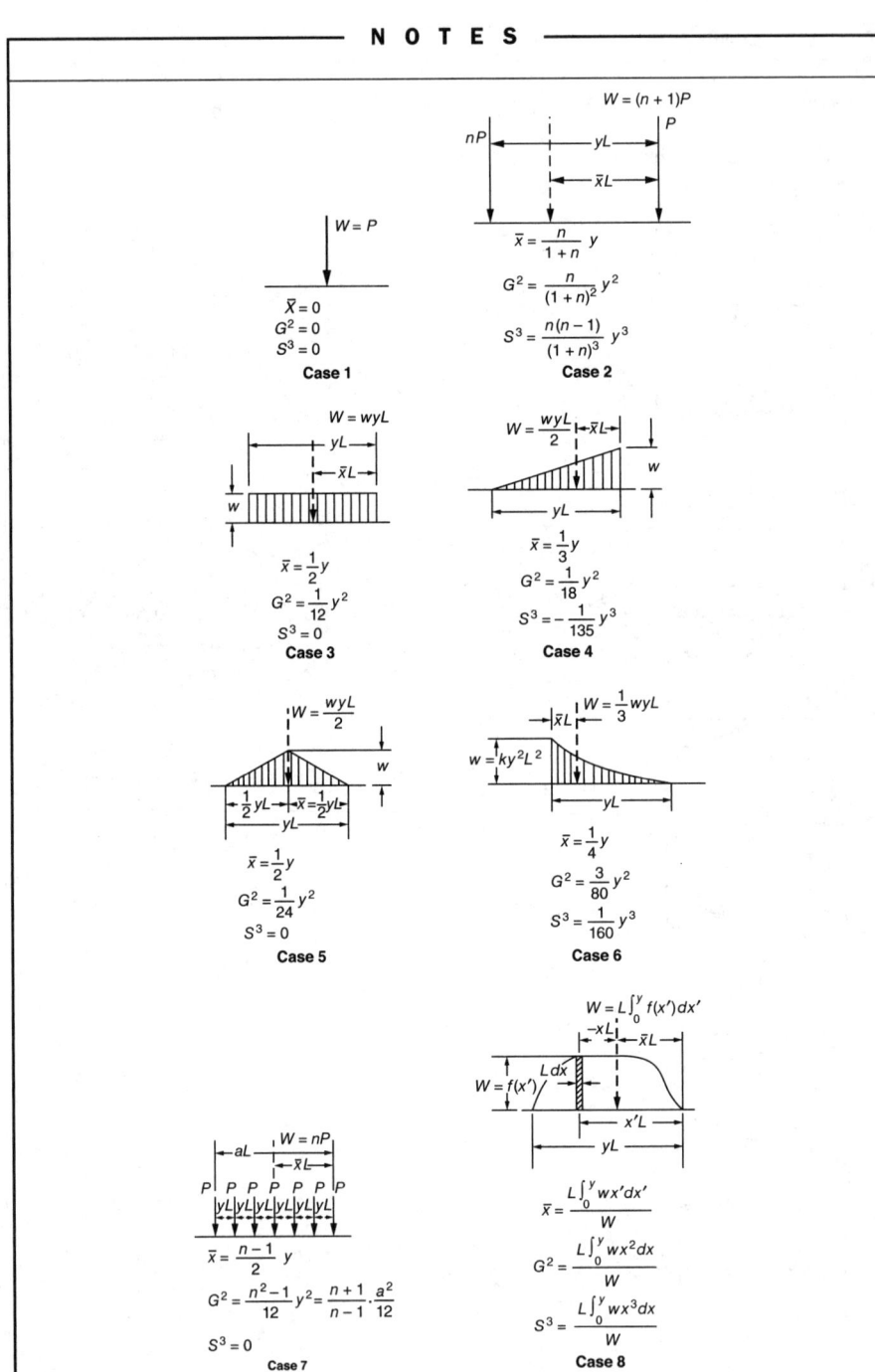

FIGURE 3.2 Characteristics of loadings.

Beams: Diagrams and Formulas for Various Loading Conditions

TABLE 3.9 Beams Fixed at Both Ends

Loadings	Support Reactions	Bending Moments (at Fixed Ends)
	$R_a = R_b = \dfrac{wc}{2}$	$M_a = M_b = -\dfrac{wcL}{24}(3-\xi^2)$ $\xi = \dfrac{c}{L}$ $M_1 = \dfrac{wcL}{4}\left(1-\dfrac{1}{2}\xi\right) - \dfrac{wcL}{24}(3-\xi^2)$ at center
	$R_a = -\dfrac{6M_0 ab}{L^3}$ $R_b = \dfrac{6M_0 ab}{L^3}$	$M_a = \dfrac{M_0 b}{L^2}(2a-b)$ $M_b = \dfrac{M_0 b}{L^2}(a-2b)$ When $x = \dfrac{L}{3}$: $M_a = 0$, $M_b = -\dfrac{M_0}{3}$
	$R_a = \dfrac{12EI}{L^3}$ $R_b = -\dfrac{12EI}{L^3}$	$M_a = -\dfrac{6EI}{L^2}$ $M_b = \dfrac{6EI}{L^2}$
	$R_a = \dfrac{6EI}{L^2}$ $R_b = -\dfrac{6EI}{L^2}$	$M_a = -\dfrac{4EI}{L}$ $M_b = \dfrac{2EI}{L}$

Continuous Beams

Continuous beams and frames are statically indeterminate. Bending moments in these beams are functions of the geometry, moments of inertia, loads, spans, and modulus of elasticity of individual members. Figure 3.3 shows how any span of a continuous beam can be treated as a single beam, with the moment diagram decomposed into basic components. Formulas for analysis are given in the diagram. Reactions of a continuous beam can be found by using the formulas in Fig. 3.4.

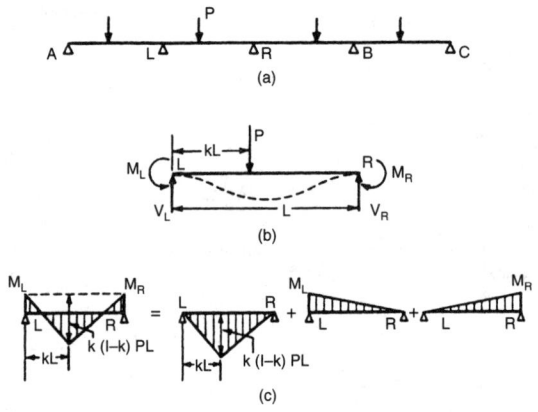

Figure 3.3 Any span of a continuous beam (a) can be treated as a simple beam, as shown in (b) and (c). In (c), the moment diagram is decomposed into basic components.

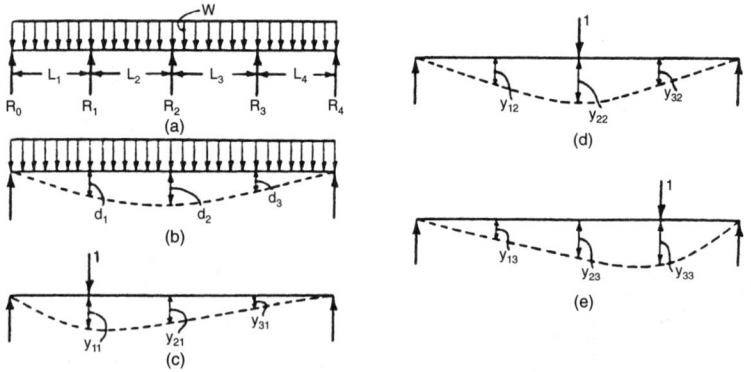

Figure 3.4 Reactions of continuous beam (a) found by making the beam statically determinate. (b) Deflections computed with interior supports removed. (c), (d), and (e) Deflections calculated for unit load over each removed support, to obtain equations for each redundant.

Beams: Diagrams and Formulas for Various Loading Conditions

TABLE 3.10 Continuous Beams

Support Reaction (R), Shear (V), Bending Moment (M), Deflection (Δ)

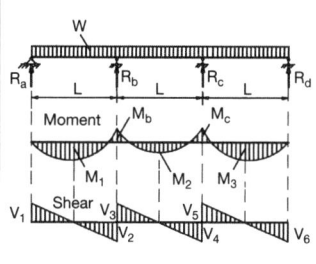

$R_a = V_1 = 0.375wL$
$R_b = V_2 + V_3 = 1.250wL, \quad V_2 = V_3 = 0.625wL$
$R_c = V_4 = 0.375wL$
$M_1 = M_2 = 0.070wL^2$, at $0.375L$ from R_a and R_c
$M_b = -0.125wL^2$
$\Delta = 0.0052 \dfrac{wL^4}{EI}$, in the middle of the spans

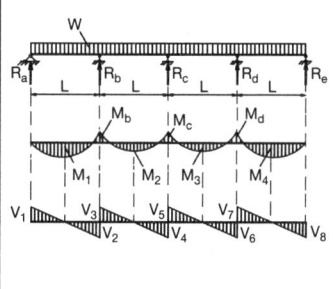

$R_a = V_1 = 0.400wL, \quad R_d = V_6 = 0.400wL$
$R_b = R_c = 1.100wL, \quad V_2 + V_5 = 0.600wL$
$V_3 = V_4 = 0.500wL$
$M_1 = M_3 = 0.080wL^2$, at $0.400L$ from R_a and R_d
$M_2 = 0.025wL^2, \quad M_b = M_c = -0.100wL^2$
$\Delta_{max} = 0.0069 \dfrac{wL^4}{EI}$, at $0.446L$ from R_a and R_d
$\Delta = 0.00675 \dfrac{wL^4}{EI}$, in the middle of spans 1 and 3
$\Delta = 0.00052 \dfrac{wL^4}{EI}$, in the middle of span 2

$R_a = R_e = 0.393wL, \quad R_b = R_d = 1.143wL$
$R_c = 0.928wL$
$V_1 = V_8 = 0.393wL, \quad V_2 = V_7 = 0.607wL$
$V_3 = V_6 = 0.536wL, \quad V_4 = V_5 = 0.464wL$
$M_1 = M_4 = 0.0772wL^2$, at $0.393L$ from R_a and R_e
$M_2 = M_3 = 0.0364wL^2$, at $0.536L$ from R_b and R_d
$M_b = M_d = -0.1071wL^2, \quad M_c = -0.0714wL^2$
$M_1 = M_4 = 0.0772wL^2$, at $0.393L$ from R_a and R_e
$\Delta_{max} = 0.0065 \dfrac{wL^4}{EI}$, at $0.440L$ from R_a and R_e

NOTES

Table 3.11 is provided for computing bending moments at the supports of elastic continuous beams with equal spans and flexural rigidity along the entire length.

The bending moments resulting from settlement of supports are added to the bending moments due to acting loads.

Example for Table 3.11. Settlement of beam support

Given. Three equal spans, continuous beam $W12 \times 35$, $L = 6.0$ m

Moment of inertia $I_z = 285 \text{ in}^4 \times 2.54^4 = 11,862.6 \text{ cm}^4$

Modulus of elasticity $E = 29,000 \text{ kip/in}^2 = \dfrac{29,000 \times 4.48222}{2.54^2}$

$= 20,147.6 \text{ kN/cm}^2$

Settlement of support B: $\Delta_B = 0.8$ cm

Required. Compute bending moments M_B and M_C.

Solution. $M_B = k_B \dfrac{EI_z}{L^2} \cdot \Delta_B = 3.6 \dfrac{20,147.6 \times 11,862.6}{(600)^2} \times 0.8$

$= 1912.0 \text{ kN} \cdot \text{cm} = 19.12 \text{ kN} \cdot \text{m}$

$M_C = k_C \dfrac{EI_z}{L^2} \cdot \Delta_B = -2.4 \dfrac{20,147.6 \times 11,862.6}{(600)^2} \times 0.8 = -1274.7 \text{ kN} \cdot \text{cm}$

$= -12.75 \text{ kN} \cdot \text{m}$

Beams: Diagrams and Formulas for Various Loading Conditions

TABLE 3.11 Continuous Beams: Settlement of Support

Bending moment at support:

$$M = k\frac{EI_z}{L^2} \cdot \Delta, \quad \text{where } k = \text{coefficient}$$

$$\Delta = \text{settlement of support}$$

Continuous Beam	Bending Moment	A	B	C	D	E	F
		\multicolumn{6}{c}{Coefficient K}					
Two Equal Spans	M_B	−1.500	3.000	−1.500			
Three Equal Spans	M_B	−1.600	3.600	−2.400	0.400		
	M_C	0.400	−2.400	3.600	−1.600		
Four Equal Spans	M_B	−1.607	3.643	−2.571	0.643	−0.107	
	M_C	0.429	−2.571	4.286	−2.571	0.429	
	M_D	−0.107	0.643	−2.571	3.643	−1.607	
Five Equal Spans	M_B	−1.608	3.645	−2.583	0.688	−0.172	0.029
	M_C	0.431	−2.584	4.335	−2.756	0.689	−0.115
	M_D	−0.115	0.689	−2.756	4.335	−2.584	0.431
	M_E	0.029	−0.172	0.688	−2.583	3.645	−1.608

Statics

NOTES

Example for Table 3.12. Moving concentrated loads

Given. Simple beam, $L = 30$ m

$P_1 = 40$ kN, $P_2 = 80$ kN, $P_3 = 120$ kN, $P_4 = 100$ kN, $P_5 = 80$ kN, $\sum P_i = 420$ kN

$a = 4$ m, $b = 3$ m, $c = 3$ m, $d = 2$ m

Required. Compute maximum bending moment and maximum end shear.

Solution. Center of gravity of loads (off load P_1):

Bending moment

$$\sum(P_i \cdot x_i) / \sum P_i = (80 \times 4 + 120 \times 7 + 100 \times 10 + 80 \times 14)/420 = 3280/420 = 7.8 \text{ m}$$

$e = 7.8 - (3 + 4) = 0.8$ m, $e/2 = 0.4$ m

$$R_A = \sum P_i \times \left(\frac{L}{2} - \frac{e}{2}\right) / L = 420(15 - 0.4)/30 = 204.4 \text{ kN}$$

$$M_{max} = R_A \cdot \left(\frac{L}{2} - \frac{e}{2}\right) - [P_1(a+b) + P_2 b] = 204.4 \times (15 - 0.4) - [40 \times (4+3) + 80 \times 3] = 2464.2 \text{ kN} \cdot \text{m}$$

End shear

Load P_1 passes off the span and P_2 moves over the left support.

$$\Delta V_1 = \frac{\sum P_i \cdot a}{L} - P_1 = \frac{420 \times 4}{30} - 40 = +16 > 0$$

Load P_2 passes off the span and P_3 moves over the left support.

$$\Delta V_2 = \frac{\sum P_i \cdot b}{L} - P_2 = \frac{420 \times 3}{30} - 80 = -38 < 0$$

For maximum end shear load P_2 is placed over the left support.

$$V_{max} = P_2 + [P_3(L-b) + P_4(L-b-c) + P_5(L-b-c-d)]/L$$

$$= 80 + [120 \times (30-3) + 100(30-3-3) + 80(30-3-3-2)]/30$$

$$= 80 + 7240/30 = 326.7 \text{ kN}$$

TABLE 3.12 Simple Beams: Moving Concentrated Loads (General Rules)

Maximum bending moment	Maximum end shear
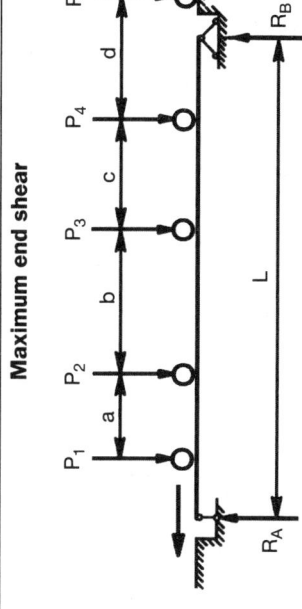	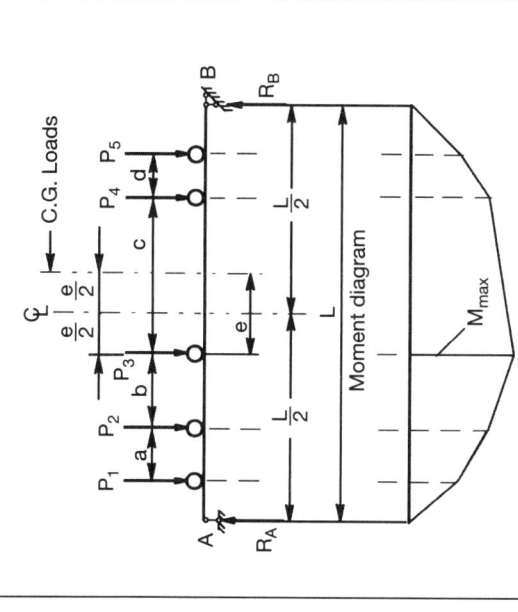
Maximum bending moment caused in a beam by a series of moving concentrated loads occurs when the center of gravity (C.G.) of all the loads and the load nearest to it (P_3 in this example) are on opposite sides of, and the same distance $\left(\dfrac{e}{2}\right)$ from, the center of the beam.	Maximum end shear in a simple beam equals the reaction when one of the moving concentrated loads is at the support. Moving loads are sequentially placed over the support, and the following expressions are evaluated: $$\Delta V_1 = \frac{\sum P \cdot a}{L} - P_1, \quad \Delta V_2 = \frac{\sum P \cdot b}{L} - P_2, \ldots$$ where $\sum P$ is the sum of the loads remaining on the beam at any time. If $\Delta V > 0$, the shear has increased. If $\Delta V > 0$, the shear has decreased. Maximum end shear occurs when the first load to produce $\Delta V < 0$ is placed over the support.

Curved Beams

The application of the flexure formula for a straight beam to the case of a curved beam results in error. When all "fibers" of a member have the same center of curvature, the *concentric* or common type of curved beam exists (Fig. 3.5). Such a beam is defined by the Winkler-Bach theory. The stress at a point y units from the centroidal axis is

$$S = \frac{M}{AR}\left[1 + \frac{y}{Z(R+y)}\right] \qquad (3.9)$$

M is the bending moment, positive when it increases curvature; y is positive when measured toward the convex side; A is the cross-sectional area; R is the radius of the centroidal axis; Z is a cross-section property defined by

$$Z = \frac{1}{A}\int \frac{y}{R+y}\,dA \qquad (3.10)$$

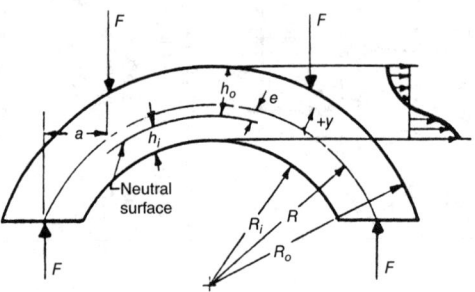

FIGURE 3.5 Curved beam.

Analytical expressions for Z of certain sections are given in Table 3.4N. Also Z can be found by *graphical* integration methods (see any advanced strength book). The neutral surface shifts toward the center of curvature, or inside fiber, an amount equal to $e = ZR/(Z + 1)$. The Winkler-Bach theory, though practically satisfactory, disregards radial stresses as well as lateral deformations and assumes pure bending. The *maximum stress* occurring on the inside fiber is $S = Mh_i/AeR_i$, whereas that on the outside fiber is $S = Mh_o/AeR_o$.

The *deflection* in curved beams can be computed by means of the moment-area theory.

The resultant deflection is then equal to $\Delta_0 = \sqrt{\Delta_x^2 + \Delta_y^2}$ in the direction defined by $\tan\theta = \Delta_y/\Delta_x$. Deflections can also be found conveniently by use of *Castigliano's theorem*. It states that in an elastic system the displacement in the direction of a force (or couple) and due to that force (or couple) is the partial derivative of the strain energy with respect to the force (or couple).

A quadrant of radius R is fixed at one end as shown in Fig. 3.6. The force F is applied in the radial direction at free end B. Then the deflection of B is by moment area.

Continued on page 74

Beams: Diagrams and Formulas for Various Loading Conditions

TABLE 3.13 Beams: Influence Lines (Examples)

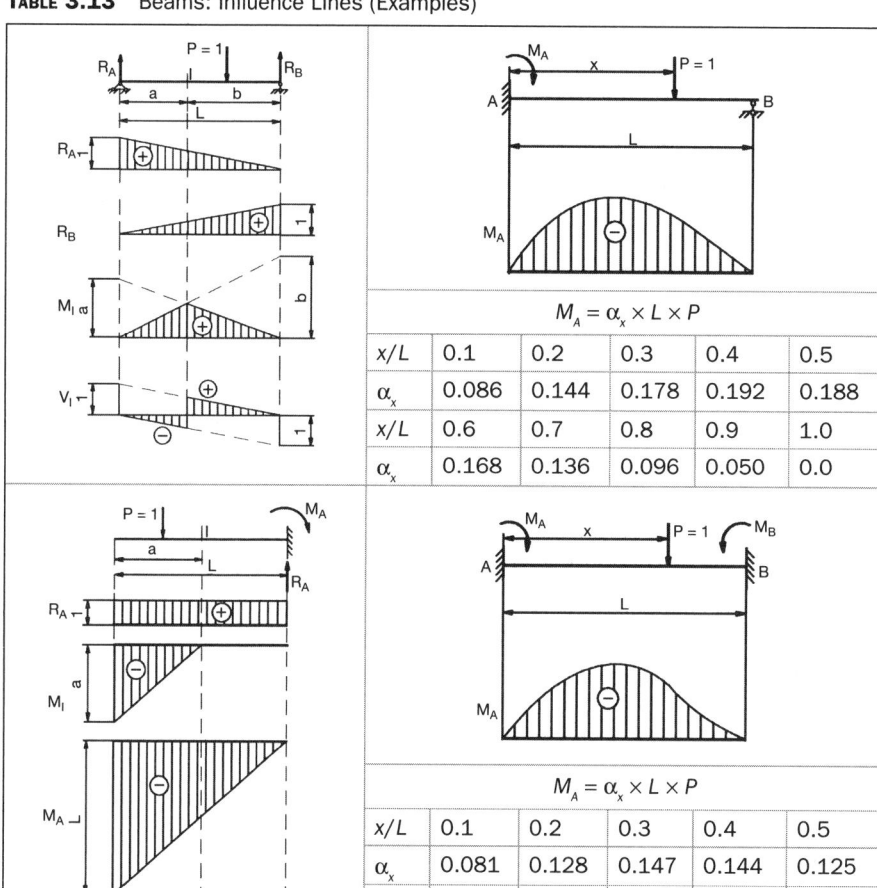

$M_A = \alpha_x \times L \times P$

x/L	0.1	0.2	0.3	0.4	0.5
α_x	0.086	0.144	0.178	0.192	0.188
x/L	0.6	0.7	0.8	0.9	1.0
α_x	0.168	0.136	0.096	0.050	0.0

$M_A = \alpha_x \times L \times P$

x/L	0.1	0.2	0.3	0.4	0.5
α_x	0.081	0.128	0.147	0.144	0.125
x/L	0.6	0.7	0.8	0.9	1.0
α_x	0.096	0.063	0.032	0.009	0.0

NOTES

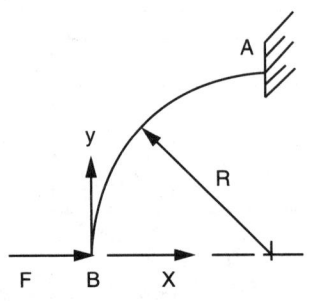

FIGURE 3.6 Quadrant with fixed end.

$$y = R\sin\theta \qquad x = R(1-\cos\theta) \qquad (3.11)$$

$$ds = R\,d\theta \qquad M = FR\sin\theta \qquad (3.12)$$

$$_B\Delta_x = \frac{\pi FR^3}{4EI} \qquad _B\Delta_y = -\frac{FR^3}{2EI} \qquad (3.13)$$

$$\text{and}\quad \Delta_B = \frac{FR^3}{2EI}\sqrt{1+\frac{\pi^2}{4}} \qquad (3.14)$$

$$\text{at}\quad \theta_x = \tan^{-1}\!\left(-\frac{FR^3}{2EI}\times\frac{4EI}{\pi FR^3}\right) \qquad (3.15)$$

$$= \tan^{-1}\frac{2}{\pi}$$

$$= 32.5°$$

By Castigliano,

$$_B\Delta_x = \frac{\pi FR^3}{4EI} \qquad _B\Delta_y = \frac{FR^3}{2EI} \qquad (3.16)$$

TABLE 3.4N Analytical Expressions for Z

Section	Expression
	$Z = -1 + \dfrac{R}{h}\!\left(\ln\dfrac{R+C}{R-C}\right)$
	$Z = -1 + 2\!\left(\dfrac{R}{r}\right)\!\left[\dfrac{R}{r} - \sqrt{\left(\dfrac{R}{r}\right)^2 - 1}\right]$
	$Z = -1 + \dfrac{R}{A}[t\,\ln(R+C_1) + (b-t)\ln(R-C_0) - b\ln(R-C_2)]$ $A = tC_1 - (b-t)C_3 + bC_2$
	$Z = -1 + \dfrac{R}{A}\!\left[b\ln\!\left(\dfrac{R+C_2}{R-C_2}\right) + (t-b)\ln\!\left(\dfrac{R+C_1}{R-C_1}\right)\right]$ $A = 2[(t-b)C_1 + bC_2]$

TABLE 3.14 Beams: Influence Lines (Examples)

Influence Lines

In studies of the variation of the effects of a moving load, such as a reaction, shear, bending moment, or stress, at a given point in a structure, use of diagrams called **influence lines** is helpful. An influence line is a diagram showing the variation of an effect as a unit load moves over a structure.

An influence line is constructed by plotting the position of the unit load as the abscissa and as the ordinate at that position, to some scale, the value of the effect being studied.

An important consequence of the reciprocal theorem is the **Mueller-Breslau principle:** The influence line of a certain effect is to some scale the deflected shape of the structure when that effect acts.

The effect, for example, may be a reaction, shear, moment, or deflection at a point. This principle is used extensively in obtaining influence lines for statically indeterminate structures.

Figure 3.7a shows the influence line for reaction at support B for a two-span continuous beam. To obtain this influence line, the support at B is replaced by a unit upward-concentrated load. The deflected shape of the beam is the influence line of the reaction at point B to some scale. To show this, let δ_{Bp} be the deflection at B due to a unit load at any point P when the support at B is removed, and let δ_{BB} be the deflection at B due to a unit load at B. Since actually reaction R_B prevents deflection at B, $R_B \delta_{BB} - \delta_{BP} = 0$. Thus $R_B = \delta_{BP}/\delta_{BB}$.

However, $\delta_{BP} = \delta_{PB}$. Hence

$$R_B = \frac{\delta_{BP}}{\delta_{BB}} = \frac{\delta_{PB}}{\delta_{BB}} \tag{3.17}$$

Similarly, influence lines may be obtained for reaction at A and moment and shear at P by the Mueller-Breslau principle, as shown in Fig. 3.7b, c, and d, respectively.

(C. H. Norris et al., *Elementary Structural Analysis;* and F. Arbabi, *Structural Analysis and Behavior,* McGraw-Hill, New York.)

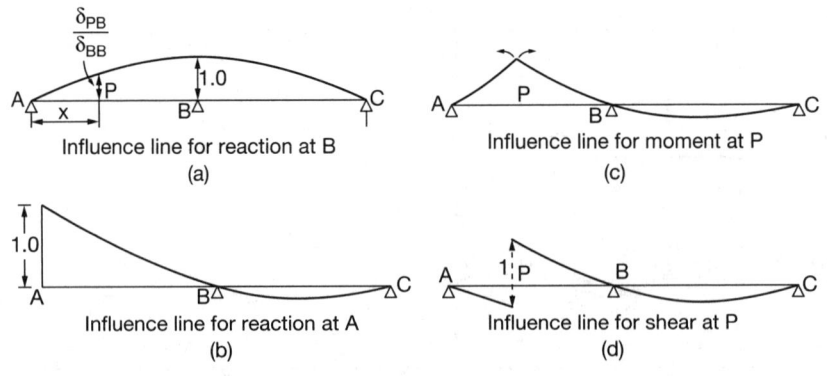

Figure 3.7 Influence lines for a two-span continuous beam.

TABLE 3.15 Beams: Computation of Bending Moment and Shear Using Influence Lines (Examples)

1. Static uniformly distributed load

For section 1–1:
Bending moment: $M_1 = w \cdot A_1$
Shear: $V_1 = w \cdot (-A_2 + A_3)$
A_1, A_2, A_3 = marked areas of influence lines

2. Static concentrated loads

For section 1–1:
Bending moment: $M_1 = \sum P \cdot h = -P_1 \cdot h_1 + P_2 \cdot h_2 + P_3 \cdot h_3 - P_4 \cdot h_4$
Shear: $V_1 = \sum P \cdot h^1 = P_1 \cdot h_1^1 - P_2 \cdot h_2^1 + P_3 \cdot h_3^1 - P_4 \cdot h_4^1$

────────── NOTES ──────────

Influence lines can be used to calculate reactions, shears, bending moments, and other effects due to fixed and moving loads. For example, Fig. 3.8a shows a simply supported beam of 60-ft span subjected to a dead load $w = 1.0$ kip/ft and a live load consisting of three concentrated loads. The reaction at A due to the dead load equals the product of the area under the influence line for the reaction at A (Fig. 3.8b) and the uniform load w. The maximum reaction at A due to the live loads may be obtained by placing the concentrated loads as shown in Fig. 3.8b and equals the sum of the products of each concentrated load and the ordinate of the influence line at the location of the load. The sum of the dead-load reaction and the maximum live-load reaction therefore is

$$R_A = \frac{1}{2} \times 1.0 \times 60 \times 1.0 + 16 \times 1.0 + 16 \times 0.767 + 4 \times 0.533 = 60.4 \text{ kips}$$

Figure 3.8c is the influence diagram for midspan bending moment with a maximum ordinate $L/4 = 60/4 = 15$. Figure 3.8c also shows the influence diagram with the live loads positioned for maximum moment at midspan. The dead-load moment at midspan is the product of w and the area under the influence line. The midspan live-load moment equals the sum of the products of each live load and the ordinate at the location of each load. The sum of the dead-load moment and the maximum live-load moment equals

$$M = \frac{1}{2} \times 15 \times 60 \times 1.0 + 16 \times 15 + 16 \times 8 + 4 \times 8 = 850 \text{ ft-kip}$$

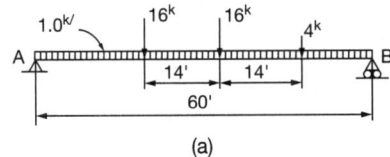

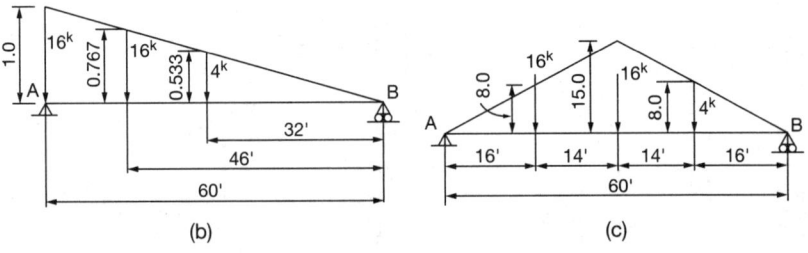

FIGURE 3.8 Determination for moving loads on a simple beam (a) of maximum end reaction (b) and maximum midspan moment (c) from influence diagrams.

TABLE 3.16 Beams: Computation of Bending Moment and Shear Using Influence Lines (Examples)

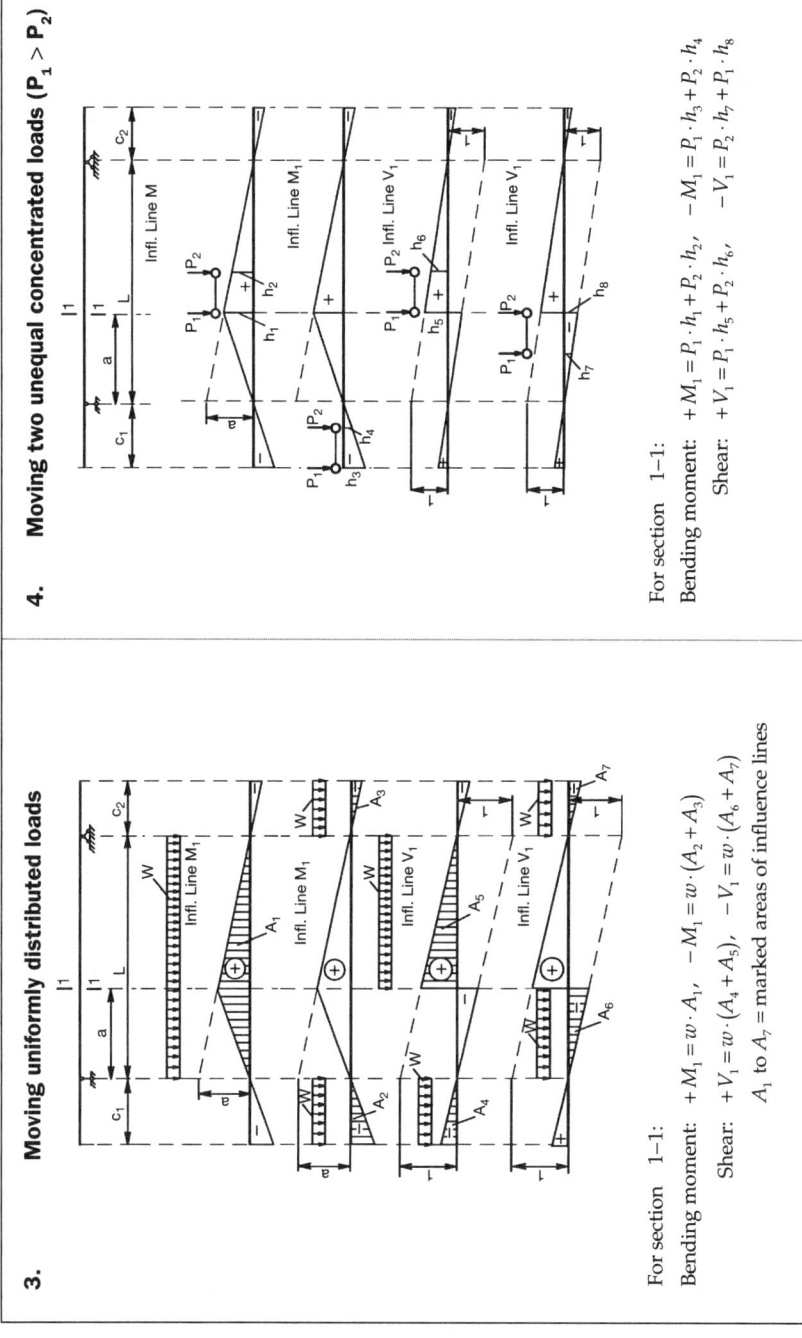

3. Moving uniformly distributed loads

For section 1–1:
Bending moment: $+M_1 = w \cdot A_1$, $\quad -M_1 = w \cdot (A_2 + A_3)$
Shear: $+V_1 = w \cdot (A_4 + A_5)$, $\quad -V_1 = w \cdot (A_6 + A_7)$
A_1 to A_7 = marked areas of influence lines

4. Moving two unequal concentrated loads ($P_1 > P_2$)

For section 1–1:
Bending moment: $+M_1 = P_1 \cdot h_1 + P_2 \cdot h_2$, $\quad -M_1 = P_1 \cdot h_3 + P_2 \cdot h_4$
Shear: $+V_1 = P_1 \cdot h_5 + P_2 \cdot h_6$, $\quad -V_1 = P_2 \cdot h_7 + P_1 \cdot h_8$

Natural Circular Frequencies and Natural Periods of Vibration of Prismatic Beams

Figure 3.9 shows the characteristic shape and gives constants for determination of natural circular frequency ω and natural period T, for the first four modes of cantilever, simply supported, fixed-end, and fixed-hinged beams. To obtain ω, select the appropriate constant from Fig. 3.9 and multiply it by $\sqrt{EI/wL^4}$. To get T, divide the appropriate constant by $\sqrt{EI/wL^4}$.

In these equations,

ω = natural frequency, rad/s
W = beam weight, lb/linear ft (kg/linear m)
L = beam length, ft (m)
E = modulus of elasticity, lb/in^2 (MPa)
I = moment of inertia of beam cross section, in^4 (mm^4)
T = natural period, s

To determine the characteristic shapes and natural periods for beams with variable cross section and mass, use the Rayleigh method. Convert the beam into a lumped-mass system by dividing the span into elements and assuming the mass of each element to be concentrated at its center. Also, compute all quantities, such as deflection and bending moment, at the center of each element. Start with an assumed characteristic shape.

Types of Support	Fundamental Mode	Second Mode	Third Mode	Fourth Mode
Cantilever $\omega\sqrt{wL^4/EI}=$ $T\sqrt{EI/wL^4}=$	20.0 0.315	0.774L 125 0.0503	0.5L 0.132L 350 0.0180	0.356L 0.094L 0.644L 684 0.0092
Simple $\omega\sqrt{wL^4/EI}=$ $T\sqrt{EI/wL^4}=$	56.0 0.112	0.5L 224 0.0281	L/3 L/3 502 0.0125	L/4 L/4 L/2 897 0.0070
Fixed $\omega\sqrt{wL^4/EI}=$ $T\sqrt{EI/wL^4}=$	127 0.0496	L/2 350 0.0180	0.359L 0.359L 684 0.0092	0.278L L/2 0.278L 1,133 0.0056
Fixed-hinged $\omega\sqrt{wL^4/EI}=$ $T\sqrt{EI/wL^4}=$	87.2 0.0722	0.56L 283 0.0222	0.384L 0.308L 591 0.0106	0.294L 0.235L 0.529L 1.111 0.0062

FIGURE 3.9 Coefficients for computing natural circular frequencies and natural periods of vibration of prismatic beams.

CHAPTER 4
Frames: Diagrams and Formulas for Various Static Loading Conditions

─── **NOTES** ───

The formulas presented in Tables 4.1 to 4.5 are used for analysis of elastic frames and allow computation of bending moments at corner sections of frame girders and posts. Bending moments at other sections of frame girders and posts can be computed using the formulas provided below.

For girders:

If $M_c > M_d$, $M_{g(x)} = M^0_{g(x)} - \left[\dfrac{M_c - M_d}{L}(L-x) + M_d \right]$

If $M_c < M_d$, $M_{g(x)} = M^0_{g(x)} - \left[\dfrac{M_d - M_c}{L}x + M_c \right]$

If $M_c = M_d = M_s$, $M_{g(x)} = M^0_{g(x)} - M_s$

For posts:

$$M_{p(x)} = M^0_{p(x)} - (H \cdot x - M_{a(b)})$$

where $M^0_{g(x)}$ and $M^0_{p(x)}$ represent, respectively, for frame girders and posts the bending moments in the corresponding simple beam due to the acting load.

Here x is the distance from the section under consideration to corner c (for the girder) and support a or b (for a post).

TABLE 4.1 Frames: Diagrams and Formulas for Various Static Loading Conditions

$$k = \frac{I_2 h}{I_1 L}$$

+M = tension on inside of frame

#	Diagram	Formulas
1	(Frame with pinned base at A, fixed at B; uniform load w on top, horizontal load H at base)	$H = \dfrac{wL^2}{4h(k+2)}$ $M_a = M_b = \dfrac{wL^2}{12(k+2)}$ $M_c = M_d = -Hh + \dfrac{wL^2}{12(k+2)}$
2	(Frame with fixed supports; uniform upward load w on base, vertical loads H_a, H_b)	$H_a = wh - H_b$ $H_b = \dfrac{wh}{8} \cdot \dfrac{2k+3}{k+2}$ $R_b = -R_a = \dfrac{wh^2}{L} \cdot \dfrac{k}{6k+1}$ $M_a = -\dfrac{wh^2}{24}\left(\dfrac{7k+15}{k+2} - \dfrac{12k}{6k+1}\right)$ $M_b = -M_a$ $M_c = H_a h - 0.5wh^2 - M_a$ $M_d = -H_b h + M_b$
3	(Frame with pinned supports at both A and B; uniform load w on top, horizontal loads H)	$H = \dfrac{wL^2}{4h(2k+3)}$ $M_c = M_d = -Hh$
4	(Frame with pin at A, fixed at B; upward uniform load w on base, vertical loads H_a, H_b)	$H_a = wh - H_b$ $H_b = \dfrac{wh}{8} \cdot \dfrac{6+5k}{2k+3}$ $M_c = H_a h - 0.5wh^2$ $M_d = -H_b h$

── NOTES ──

Example for Table 4.5. Analysis of frame

Given. Frame 5 in Table 4.5, $L = 12$ m, $h = 3$ m

Posts $W10 \times 45$, $I_1 = 248$ in$^4 \times 2.54^4 = 10,322$ cm^4

Girder $W14 \times 82$, $I_2 = 882$ in$^4 \times 2.54^4 = 36,712$ cm^4

Load $P = 20$ kN, $a = 4$ m, $b = 8$ m

Required. Compute support reactions and bending moments.

Solution. $k = \dfrac{I_2 h}{I_1 L} = \dfrac{36,712 \times 3}{10,322 \times 12} = 0.889$, $\xi = \dfrac{a}{L} = \dfrac{4}{12} = 0.333$

$$H = \frac{3}{2} \cdot \frac{Pab}{hL(k+2)} = \frac{3}{2} \cdot \frac{20 \times 4 \times 8}{3 \times 12(0.889 + 2)} = 9.23 \text{ kN}$$

$$R_a = \frac{Pb}{L} \cdot \frac{1 + \xi - 2\xi^2 + 6k}{6k + 1} = 13.57 \text{ kN}$$

$$R_b = P - R_a = 20 - 13.57 = 6.43 \text{ kN}$$

$$M_a = \frac{Pab}{2L} \cdot \frac{5k - 1 + 2\xi(k+2)}{(k+2)(6k+1)} = 7.813 \text{ kN} \cdot \text{m}$$

$M_b = R_a L + M_a - Pb = 13.57 \times 12 + 7.813 - 20 \times 8 = 10.653$ kN·m

$M_c = -Hh + M_a = -9.23 \times 3 + 7.813 = -19.877$ kN·m

$M_d = -Hh + M_b = -9.23 \times 3 + 10.653 = -17.037$ kN·m

Bending moment at point of load

$$M_g = M_g^0 - \left[\frac{M_c - M_d}{L}(L - a) + M_d \right], \quad M_g^0 = \frac{Pab}{L}$$

$$M_g = \frac{20 \times 4 \times 8}{12} - \left[\frac{19.877 - 17.037}{12}(12 - 4) + 17.037 \right] = 34.403 \text{ kN} \cdot \text{m}$$

TABLE 4.2 Frames: Diagrams and Formulas for Various Static Loading Conditions

5

$$H = \frac{3}{2} \cdot \frac{Pab}{hL(k+2)}, \quad \xi = \frac{a}{L}$$

$$R_a = \frac{Pb}{L} \cdot \frac{1+\xi-2\xi^2+6k}{6k+1}$$

$$R_b = P - R_a$$

$$M_a = \frac{Pab}{2L} \cdot \frac{5k-1+2\xi(k+2)}{(k+2)(6k+1)}$$

$$M_b = M_a + R_a L - Pb$$

$$M_c = -Hh + M_a$$

$$M_d = -Hh + M_b$$

7

$$H = \frac{3}{2} \cdot \frac{Pab}{hL(2k+3)}$$

$$M_c = M_d = -Hh$$

6

$$H = \frac{P}{2}$$

$$R_a = -R_b = -\frac{3Ph}{L} \cdot \frac{k}{6k+1}$$

$$M_a = -M_b = -\frac{Ph}{2} \cdot \frac{3k+1}{6k+1}$$

$$M_c = Hh - M_a$$

$$M_d = -Hh + M_b$$

8

$$H_a = H_b = \frac{P}{2}$$

$$M_c = -M_d = \frac{Ph}{2}$$

Columns and Frames

Columns are structural members subjected to direct compression. All columns can be grouped into the following three classes:

1. *Compression blocks* are so short (with a slenderness ratio—that is, unsupported length divided by the lowest radius of gyration of the member—below 30) that bending is not potentially occurring.
2. Columns so slender that bending under load is given are termed *long Columns* and are defined by Euler's theory.
3. Intermediate-length columns, often used in structural practice, are called *short Columns*.

Columns are widely used in a variety of frames.

Long and short columns usually fail by buckling when their *critical load* is reached. Long columns are analyzed using Euler's column formula, namely,

$$P_{cr} = \frac{n\pi^2 EI}{l^2} = \frac{n\pi^2 EA}{(l/r)^2} \qquad (4.1)$$

In this formula, the coefficient n accounts for end conditions. When the column is pivoted at both ends, $n = 1$; when one end is fixed and the other end is rounded, $n = 2$; when both ends are fixed, $n = 4$; and when one end is fixed and the other is free, $n = 0.25$. The slenderness ratio separating long columns from short columns depends on the modulus of elasticity and the yield strength of the column material. When Euler's formula results in $(P_{cr}/A) > S_y$, strength instead of buckling causes failure, and the column ceases to be long. In quick estimating numbers, this *critical slenderness ratio* falls between 120 and 150.

TABLE 4.3 Frames: Diagrams and Formulas for Various Static Loading Conditions

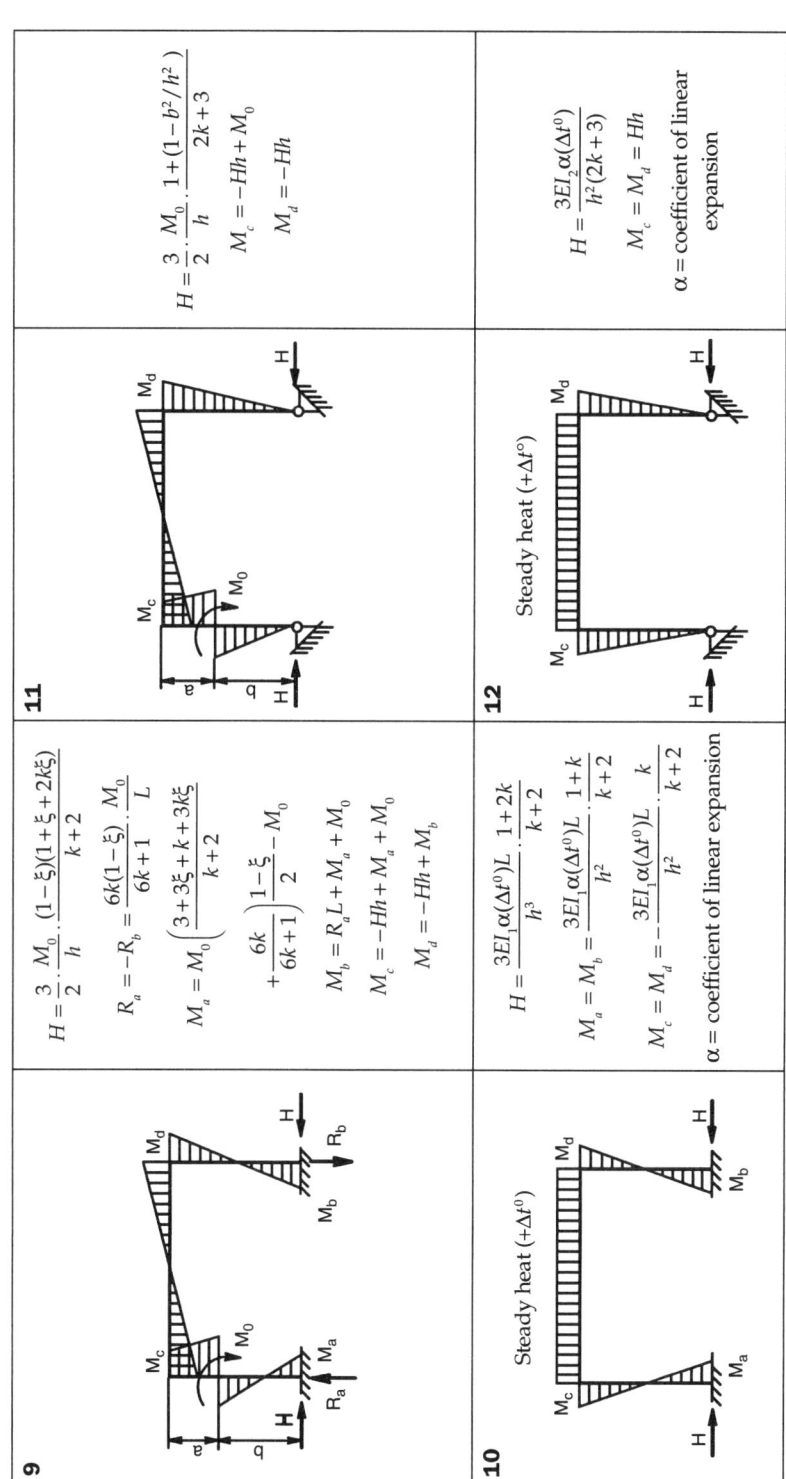

9

$$H = \frac{3}{2} \cdot \frac{M_0}{h} \cdot \frac{(1-\xi)(1+\xi+2k\xi)}{k+2}$$

$$R_a = -R_b = \frac{6k(1-\xi)}{6k+1} \cdot \frac{M_0}{L}$$

$$M_a = M_0 \left(\frac{3+3\xi+k+3k\xi}{k+2} \right.$$
$$\left. + \frac{6k}{6k+1} \right) \frac{1-\xi}{2} - M_0$$

$$M_b = R_a L + M_a + M_0$$

$$M_c = -Hh + M_a + M_0$$

$$M_d = -Hh + M_b$$

10

Steady heat $(+\Delta t°)$

$$H = \frac{3EI_1\alpha(\Delta t°)L}{h^3} \cdot \frac{1+2k}{k+2}$$

$$M_a = M_b = \frac{3EI_1\alpha(\Delta t°)L}{h^2} \cdot \frac{1+k}{k+2}$$

$$M_c = M_d = -\frac{3EI_1\alpha(\Delta t°)L}{h^2} \cdot \frac{k}{k+2}$$

α = coefficient of linear expansion

11

$$H = \frac{3}{2} \cdot \frac{M_0}{h} \cdot \frac{1+(1-b^2/h^2)}{2k+3} + M_0$$

$$M_c = -Hh + M_0$$

$$M_d = -Hh$$

12

Steady heat $(+\Delta t°)$

$$H = \frac{3EI_2\alpha(\Delta t°)}{h^2(2k+3)}$$

$$M_c = M_d = Hh$$

α = coefficient of linear expansion

— NOTES —

Short Columns

Stress in short columns can be considered to be partly due to compression and partly due to bending. Empirical, rational expressions for column stress are, in general, based on the assumption that the permissible stress must be reduced below that which could be permitted were it due to compression only. The manner in which this reduction is made determines the type of equation and the slenderness ratio beyond which the equation does not apply. Typical column formulas are given in Table 4.1N.

TABLE 4.1N Typical Short-Column Formulas

Formula	Material	Code	Slenderness Ratio
$S_w = 17{,}000 - 0.485\left(\dfrac{l}{r}\right)^2$	Carbon steels	AISC	$\dfrac{l}{r} < 120$
$S_w = 16{,}000 - 70\left(\dfrac{l}{r}\right)$	Carbon steels	Chicago	$\dfrac{l}{r} < 120$
$S_w = 15{,}000 - 50\left(\dfrac{l}{r}\right)$	Carbon steels	AREA	$\dfrac{l}{r} < 150$
$S_w = 19{,}000 - 100\left(\dfrac{l}{r}\right)$	Carbon steels	Am.Br.Co.	$60 < \dfrac{l}{r} < 120$
$^*S_{cr} = 135{,}000 - \dfrac{15.9}{c}\left(\dfrac{l}{r}\right)^2$	Alloy-steel tubing	ANC	$\dfrac{l}{\sqrt{cr}} < 65$
$S_w = 9{,}000 - 40\left(\dfrac{l}{r}\right)$	Cast iron	NYC	$\dfrac{l}{r} < 70$
$^*S_{cr} = 34{,}500 - \dfrac{245}{\sqrt{c}}\left(\dfrac{l}{r}\right)$	2017ST aluminum	ANC	$\dfrac{l}{\sqrt{cr}} < 94$
$^*S_{cr} = 5{,}000 - \dfrac{0.5}{c}\left(\dfrac{l}{r}\right)^2$	Spruce	ANC	$\dfrac{l}{\sqrt{cr}} < 72$
$^*S_{cr} = S_y\left[1 - \dfrac{S_y}{4n\pi^2 E}\left(\dfrac{1}{r}\right)^2\right]$	Steels	Johnson	$\dfrac{l}{r} < \sqrt{\dfrac{2n\pi^2 E}{S_y}}$
$^\dagger S_{cr} = \dfrac{S_y}{1 + \dfrac{ec}{r^2}\sec\left(\dfrac{l}{r}\sqrt{\dfrac{P}{4AE}}\right)}$	Steels	Secant	$\dfrac{l}{r} < $ critical

$^*S_{cr}$ = theoretical maximum; c = end fixity coefficient; $c = 2$, both ends pivoted; $c = 2.86$, one end pivoted, other end fixed; $c = 4$, both ends fixed; $c = 1$ one end fixed, one end free.
† Initial eccentricity at which load is applied to center of column cross section.

TABLE 4.4 Frames: Diagrams and Formulas for Various Static Loading Conditions

	$I_2 = \infty$		$I_2 = \infty$
			$n = \dfrac{I_t}{I_b}, \quad \lambda = \dfrac{a}{h}$
			$\delta_{11} = \left(1 - \lambda^3 + \dfrac{\lambda^3}{n}\right)\dfrac{2h^3}{3EI_b}$
13	$S = \dfrac{3}{6}wh, \quad H_a = \dfrac{13}{16}wh$ $M_a = -\dfrac{5}{16}wh^2, \quad M_b = \dfrac{3}{16}wh^2$	**16**	$S = \dfrac{\delta_{1w}}{\delta_{11}}, \quad H_b = -S$ $M_a = \dfrac{wh^2}{2} - Sh, \quad M_b = Sh$ $\delta_{1w} = \left(1 - \lambda^4 + \dfrac{\lambda^4}{n}\right)\dfrac{wh^4}{8EI_b}$
14	$S = \dfrac{P}{2}, \quad H_a = H_b = \dfrac{P}{2}$ $M_a = -M_b = -\dfrac{Ph}{2}$	**17**	$S = \dfrac{P}{2}, \quad H_a = H_b = \dfrac{P}{2}$ $M_a = -M_b = -\dfrac{Ph}{2}$
15	$S = 0.75(1 - \lambda^2)\dfrac{M_0}{h}$ $H = -S$ $M_a = Sh - M_0, \quad M_b = Sh$	**18**	$S = \dfrac{\delta_{1m}}{\delta_{11}}, \quad H = -S$ $M_a = Sh - M_0, \quad M_b = Sh$ $\delta_{1m} = (1 - \lambda^2)\dfrac{M_0 h^2}{2EI_b}$

Elastic Flexural Buckling of Columns

Elastic buckling is a state of lateral instability that occurs while the material is stressed below the yield point. It is of special importance in structure with slender members. Euler's formula for pin-ended columns (Fig. 4.1) gives valid results for the critical buckling load, kip (N). This formula is, with L/r as the slenderness ratio of the column.

$$P = \frac{\pi^2 EA}{(L/r)^2} \tag{4.1}$$

where E = modulus of elasticity of column material, psi (MPa)
A = column cross-sectional area, in² (mm²)
r = radius of gyration of column, in (mm)

Figure 4.2 show some ideal end conditions for slender columns and corresponding crictical buckling load. Elastic critical buckling loads may be obtained for all cases by substituting an effective length KL for the length L of the pinned column, giving

$$P = \frac{\pi^2 EA}{(KL/r)^2} \tag{4.2}$$

In some cases of columns with open sections, such as a cruciform section, the controlling buckling mode may be one of twisting instead of lateral deformation. If the warping rigidity of the section is negligible, *torsional buckling* in a pin-ended column occurs at an axial load of

$$P = \frac{GJA}{I_p} \tag{4.3}$$

where G = shear modulus of elasticity, psi (MPa)
J = torsional constant
A = cross-sectional area, in² (mm²)
I_p = polar moment of inertia = $I_x + I_y$, in⁴ (mm⁴)

If the section possesses a significant amount of warping rigidity, the axial buckling load is increased to

$$P = \frac{A}{I_p}\left(GJ + \frac{\pi^2 EC_w}{L^2}\right) \tag{4.4}$$

where C_w is the warping constant, a function of cross-sectional shape and dimensions. Figures 4.1 and 4.2 show buckling in columns.

Continued on page 92

Frames: Diagrams and Formulas for Various Static Loading Conditions

TABLE 4.5 Frames: Diagrams and Formulas for Various Static Loading Conditions

19 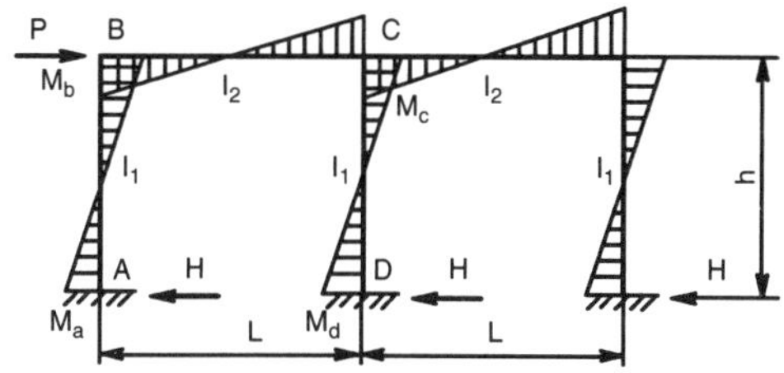 *[frame diagram with P, Mb, B, C, I2, Mc, I1, h, A, H, Ma, D, Md, L]*	$M_a \approx -\dfrac{Ph}{6}, \quad M_b \approx +\dfrac{Ph}{6}$ $M_c \approx +\dfrac{Ph}{6}, \quad M_d \approx -\dfrac{Ph}{6}$
20 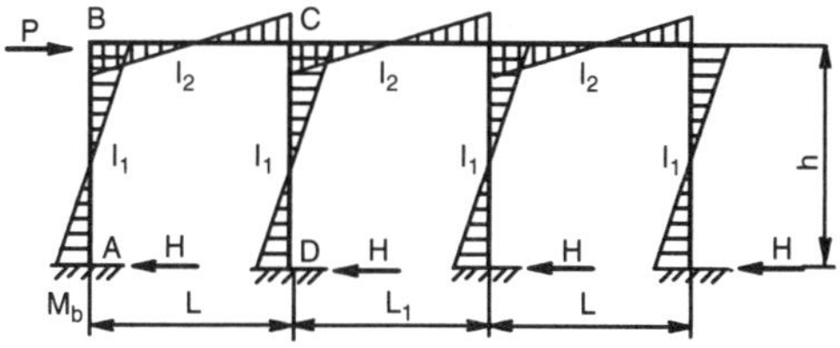 *[frame diagram with P, B, C, I2, I1, h, A, H, D, Mb, L, L1]*	$M_a \approx -\dfrac{Ph}{8}, \quad M_b \approx +\dfrac{Ph}{8}$ $M_c \approx +\dfrac{Ph}{8}, \quad M_d \approx -\dfrac{Ph}{8}$
21 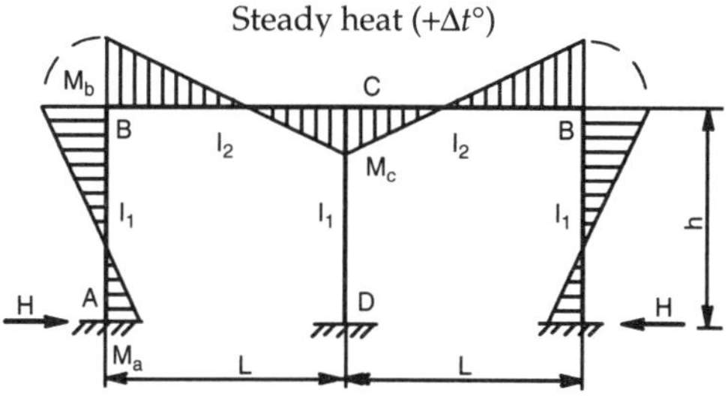 *[frame diagram, Steady heat (+Δt°), Mb, B, C, I2, Mc, I1, h, H, A, Ma, D, L]*	$M_a = \dfrac{3EI_1(2k+1)}{h^2(1+k)}\,\alpha \cdot \Delta t^\circ L$ $M_b = -\dfrac{6EI_1 k}{h^2(1+k)}\,\alpha \cdot \Delta t^\circ L$ $M_c = -\dfrac{1}{2}M_b, \quad k = \dfrac{I_2 h}{I_1 L}$ α = coefficient of linear expansion
22 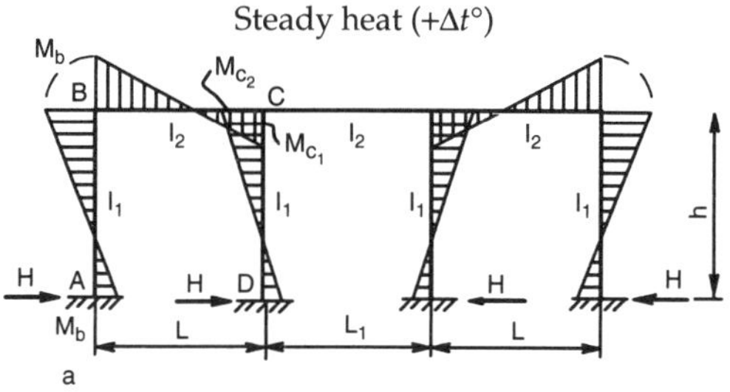 *[frame diagram, Steady heat (+Δt°), Mb, B, Mc2, C, Mc1, I2, I1, h, H, A, D, Mb, L, L1, a]*	$M_a = \dfrac{3EI_1(2k+1)}{h^2(1+k)}\left(L + \dfrac{L_1}{2}\right)\alpha \cdot \Delta t^\circ$ $M_b = -\dfrac{6EI_1 k}{h^2(1+k)}\left(L + \dfrac{L_1}{2}\right)\alpha \cdot \Delta t^\circ$ $M_{c1} = -\dfrac{1}{2}M_b, \quad M_{c2} = -\dfrac{6EI_1}{h^2}\left(\dfrac{L_1}{2}\right)\alpha \Delta t^\circ$ $M_d = -M_{c2}, \quad k = \dfrac{I_2 h}{I_1 L}$ α = coefficient of linear expansion

NOTES

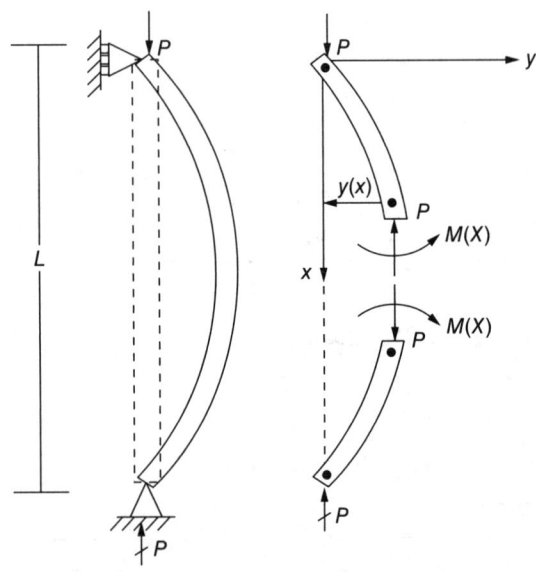

FIGURE 4.1 (a) Buckling of a pin-ended column under axial load. (b) Internal forces hold the column in equilibrium.

Type of Column	Effective Length	Critical Buckling Load
(pinned-pinned)	L	$\dfrac{\pi^2 EI}{L^2}$
(fixed-fixed)	$\dfrac{L}{2}$	$\dfrac{4\pi^2 EI}{L^2}$
(fixed-pinned)	$\approx 0.7L$	$\approx \dfrac{2\pi^2 EI}{L^2}$
(fixed-free)	$2L$	$\dfrac{\pi^2 EI}{4L^2}$

FIGURE 4.2 Buckling formulas for columns.

CHAPTER 5
Arches: Diagrams and Formulas for Various Loading Conditions

--- **N O T E S** ---

Tables 5.1 to 5.9 are provided for determining support reactions and bending moments in elastic arches with constant or variable cross sections.

Table 5.1 includes formulas for computing in any cross section k the axis force N_k and the shear V_k. These formulas can also be applied in the analysis of arches shown in Tables 5.2 to 5.9.

Bending moment $\quad M_k = R_A \cdot x_k - H_A \cdot y_k \pm M_A - \sum_{\text{left}} P_i \cdot a_i$

Axial force $\quad N_k = R_A \sin\phi + H_A \cos\phi - \sum_{\text{left}} P_i \sin\phi$

Shear $\quad V_k = R_A \cos\phi - H_A \sin\phi - \sum_{\text{left}} P_i \cos\phi$

where a_i = distance from load P to point k

TABLE 5.1 Three-Hinged Arches: Support Reactions, Bending Moment, and Axial Force

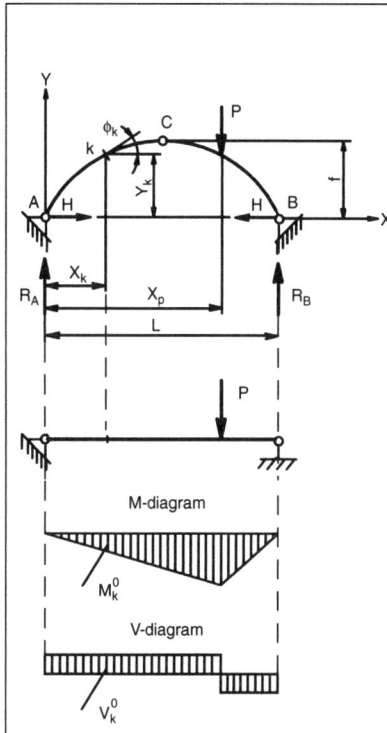

Vertical reactions:
$$\sum M_B = R_A L - P(L - x_P) = 0, \quad R_A = P\frac{L - x_P}{L};$$
$$\sum M_A = -R_B L + P x_P = 0, \quad R_B = P\frac{x_P}{L}.$$

Horizontal reactions:
$$\sum M_C = R_A \frac{L}{2} - H_A f = 0, \quad H_A = R_A \frac{L}{2f};$$
$$\sum X = H_A - H_B = 0, \quad H_B = H_A = H.$$

Section $k\,(x_k, y_k)$

Bending moment: $M_k = \sum_{\text{left}} M = R_A x_k - H y_k,$

or $M_k = M_k^0 - H y_k.$

Shear: $V_k = \left(R_A - \sum_{\text{left}} P\right) \cos\phi_k - H \sin\phi_k,$

or $V_k = V_k^0 \cos\phi_k - H \sin\phi_k.$

Axial force: $N_k = \left(R_A - \sum_{\text{left}} P\right) \sin\phi_k + H \cos\phi_k,$

or $N_k = V_k^0 \sin\phi_k + H \cos\phi_k.$

M_k^0 and V_k^0 = bending moment and shear in simple beam for section x_k

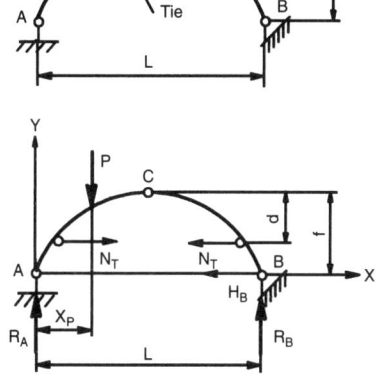

Vertical reactions:
$$\sum M_B = R_A L - P(L - x_P) = 0, \quad R_A = P\frac{L - x_P}{L};$$
$$\sum M_A = -R_B L + P x_P = 0, \quad R_B = P\frac{x_P}{L}.$$

Horizontal reaction:
$$\sum X = -H_B = 0.$$

Force N_T:
$$\sum_{\text{left}} M_C = R_A \frac{L}{2} - N_T d - P\left(\frac{L}{2} - x_P\right) = 0,$$

$$N_T = \frac{1}{d}\left[P\left(\frac{L}{2} - x_P\right) - R_A \frac{L}{2}\right],$$

or $\sum_{\text{right}} M_C = N_T d - R_B \frac{L}{2} = 0, \quad N_T = R_b \frac{L}{2d}.$

Example for Table 5.2. Symmetrical three-hinged arch

Given. Circular arch 2 in Table 5.2, $L = 20$ m, $f = 4$ m

$$\text{Radius } R = \frac{4f^2 + L^2}{8f} = \frac{4 \times 4^2 + 20^2}{8 \times 4} = 14.5 \text{ m}, \quad x_m = 5 \text{ m}$$

$$y_m = \sqrt{R^2 - \left(\frac{L}{2} - x_m\right)^2} - (R - f) = \sqrt{14.5^2 - (10 - 5)^2} - (14.5 - 4) = 3.11 \text{ m}$$

$$\tan \phi_m = \left(\frac{L}{2} - x_m\right)/(R - f + y_m) = (10 - 5)/(14.5 - 4 + 3.11) = 0.367$$

$\phi_m = 20.17°$, $\sin \phi_m = 0.345$, $\cos \phi_m = 0.939$

Distribution load $w = 2$ kN/m

Required. Compute support reactions R_A and H_A, support bending moment M_A, bending moment M_m, axial force N_m, and shear V_m.

Solution. $R_A = \frac{3}{8}wL = \frac{3}{8} \times 2 \times 20 = 15$ kN, $H_A = \frac{wL^2}{16f} = \frac{2 \times 20^2}{16 \times 4} = 12.5$ kN

$$\xi_m = \frac{x_m}{L} = \frac{5}{20} = 0.25, \quad \eta_m = \frac{y_m}{f} = \frac{3.11}{4} = 0.778$$

$$M_m = \frac{wL^2}{16}\left[8(\xi_m - \xi_m^2) - 2\xi_m - \eta_m\right] = \frac{2 \times 20^2}{16}\left[8(0.25 - 0.25^2) - 2 \times 0.25 - 0.778\right] = 11.1 \text{ kN} \cdot \text{m}$$

$N_m = R_A \sin\phi_m + H_A \cos\phi_m - w \cdot x_m \sin\phi_m = 15 \times 0.345 + 12.5 \times 0.939 - 2 \times 5 \times 0.345 = 13.46$ kN

$V_m = R_A \cos\phi_m - H_A \sin\phi_m - w \cdot x_m \cos\phi_m = 15 \times 0.939 - 12.5 \times 0.345 - 2 \times 5 \times 0.939 = 0.38$ kN

Arches: Diagrams and Formulas for Various Loading Conditions

TABLE 5.2 Symmetrical Three-Hinged Arches of Any Shape: Formulas for Various Static Loading Conditions

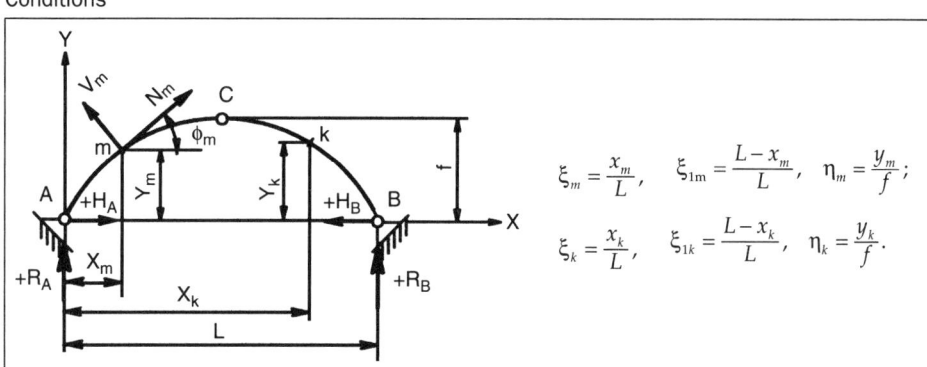

$$\xi_m = \frac{x_m}{L}, \quad \xi_{1m} = \frac{L-x_m}{L}, \quad \eta_m = \frac{y_m}{f};$$

$$\xi_k = \frac{x_k}{L}, \quad \xi_{1k} = \frac{L-x_k}{L}, \quad \eta_k = \frac{y_k}{f}.$$

Loadings	Support Reactions	Bending Moments
1. 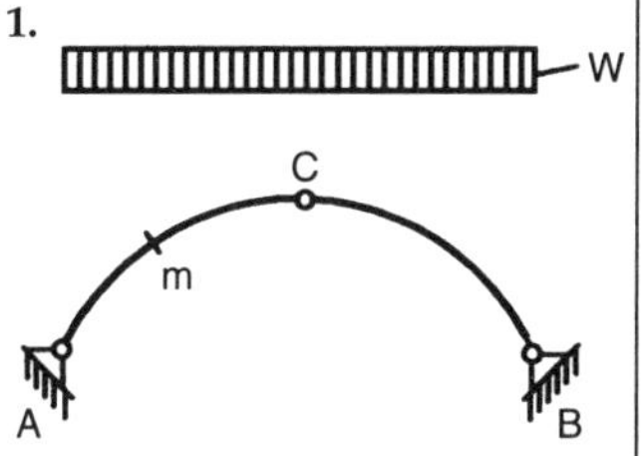 (uniform load w over full span)	$R_A = R_B = \dfrac{wL}{2},$ $H_A = H_B = \dfrac{wL^2}{8f}.$	$M_m = \dfrac{wL^2}{8}\left[4(\xi_m - \xi_m^2) - \eta_m\right].$
2. (uniform load w over left half $L/2$) 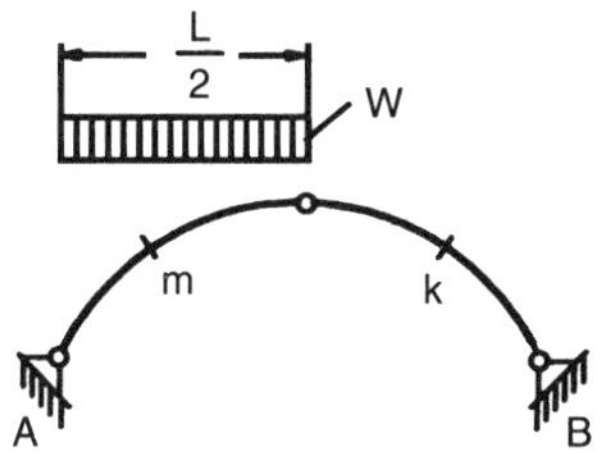	$R_A = \dfrac{3}{8}wL, \quad R_B = \dfrac{1}{8}wL,$ $H_A = H_B = \dfrac{wL^2}{16f}.$	$M_m = \dfrac{wL^2}{16}\left[8(\xi_m - \xi_m^2) - 2\xi_m - \eta_m\right],$ $M_k = \dfrac{wL^2}{16}(2\xi_k - \eta_k).$
3. 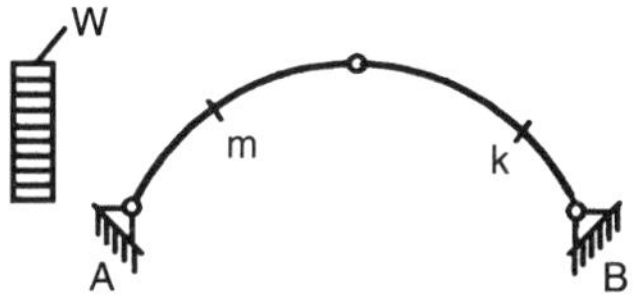 (triangular load w on left side)	$R_A = -\dfrac{wf^2}{2L}, \quad R_B = \dfrac{wf^2}{2L},$ $H_A = -\dfrac{3}{4}wf,$ $H_B = \dfrac{1}{4}wf.$	$M_m = -\dfrac{wf^2}{2}\left(\xi_m - \dfrac{3}{2}\eta_m + \eta_m^2\right),$ $M_k = \dfrac{wf^2}{4}(2\xi_{1k} - \eta_k).$
4. 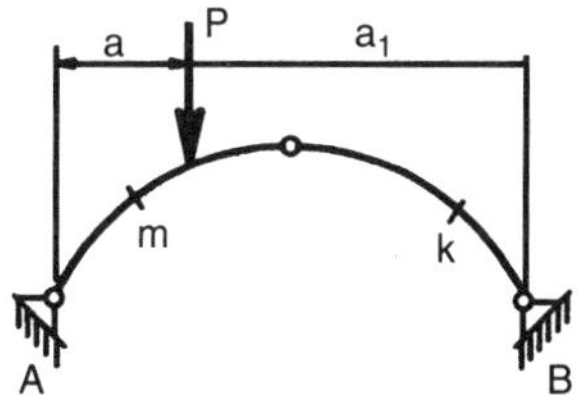 (point load P at distance a from A, a_1 from B)	$R_A = P\dfrac{a_1}{L}, \quad R_B = P\dfrac{a}{L},$ $H_A = H_B = P\dfrac{a}{2f}.$	$M_m = P\dfrac{a}{2}\left(2\dfrac{a_1}{a}\xi_m - \eta_m\right),$ $M_k = P\dfrac{a}{2}(2\xi_{1k} - \eta_k).$

NOTES

TABLE 5.1N Formulas for Circular Rings and Arches*

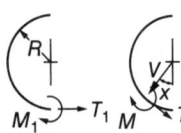

M_1, T_1, V_1, M, T, and V are positive when as shown, negative when reversed. All applied forces and couples are positive when as shown, negative when reversed. The following notation is employed: E = modulus of elasticity (lb per sq in); I = moment of inertia of ring cross section (in^4); W or F as shown = applied load or reaction (lb); w = applied load (lb per lin in); k = weight of contained liquid (lb per cu in); $z = \sin x$, $u = \cos x$; $s = \sin \theta$, $c = \cos \theta$; $n = \sin \phi$, $e = \cos \phi$; $p = \sin \beta$, $q = \cos \beta$. All angles in radians, distances in inches, forces in pounds, and moments in inch-pounds. $+D_x$ or $+D_y$ means increase, $-D_x$ or $-D_y$ means decrease in diameters. $+\Delta R$ means increase, $-\Delta R$ means decrease, in upper half of vertical diameter.

Loading, Support, and Case Number	Formulas for End Reactions — Circular Arch	Parabolic Arch; l = Span; f = Rise; a = Distance Load from Left End, b from Right
1. Ends pinned: concentrated load W at any point ϕ	$H = \dfrac{1}{2}W\left[\dfrac{s^2 - n^2 - 2c(\theta s - \phi n + c - e) - \alpha(s^2 - n^2)}{\theta - 3sc + 2\theta c^2 + \alpha(\theta + sc)}\right]$ $V_1 = \dfrac{1}{2}W\left(\dfrac{s+n}{s}\right)$ (Here $\alpha = \dfrac{I}{AR^2}$ where A = cross-sectional area)	$H = W\dfrac{5b}{8f}\left[1 - 2\left(\dfrac{b}{l}\right)^2 + \left(\dfrac{b}{l}\right)^3\right]$ $V_1 = W\dfrac{b}{l}$ (Ref. 1)
2. Like Case 1 except ends fixed	$H = \dfrac{1}{2}W\left[\dfrac{\dfrac{2}{\theta}(se - \phi sn - sc) - s^2 - n^2 - \alpha(s^2 - n^2)}{\theta + sc - \dfrac{2s^2}{\theta} + \alpha(\theta + sc)}\right]$ $V_1 = \dfrac{1}{2}W\left(\dfrac{\theta + \phi - cs + en - 2nc}{\theta - cs}\right)$ $M_1 = V_1 R s + HR\left(\dfrac{\theta c - s}{\theta}\right) + \dfrac{1}{2}WR\left(\dfrac{c - e - \phi n - \theta n}{\theta}\right)$ (α as for Case 27)	$H = \dfrac{15W}{4}\left(\dfrac{a^2 b^2}{l^3 f}\right)$ $V_1 = W\left[1 - \left(\dfrac{a}{l}\right)^2 - \dfrac{2a^2 b}{l^3}\right]$ $M_1 = Wa\left(\dfrac{5ab^2}{2l^3} - \dfrac{b^2}{l^2}\right)$ (Ref. 1)
3. Ends pinned: uniform load w lb per linear in	$H = \dfrac{1}{2}wR\left[\dfrac{\dfrac{1}{3}s^3 + \phi c - 2\theta s^2 c - sc^2 + 2\alpha\left(\theta c^2 - \dfrac{1}{2}\theta - s\dfrac{1}{2}sc\right)}{2\theta c^2 + \theta - 3sc + \alpha(\theta + sc)}\right]$ $V_1 = wRs$ (α as for Case 1)	$H = \dfrac{wl^2}{8f}$ $V_1 = \dfrac{1}{2}wl$ (Ref. 1)
4. Like Case 3 except ends fixed	$H = wR\left[\dfrac{\dfrac{1}{4}\left(\dfrac{s^2 c}{\theta} - s\right) + \dfrac{1}{6}s^3 + \alpha\left(\dfrac{1}{2}\theta - \dfrac{1}{2}\theta c^2 + \dfrac{1}{4}sc\right)}{\dfrac{(\theta - s)^2}{\theta} - \dfrac{3}{2}\theta + 2s - \dfrac{1}{2}sc - \alpha\left(\dfrac{1}{2}\theta + \dfrac{1}{2}sc\right)}\right]$ $M_1 = wR^2\left(\dfrac{1}{2}s^2 - \dfrac{1}{4} + \dfrac{1}{4}\dfrac{sc}{\theta}\right) - HR\left(\dfrac{s}{\theta} - c\right)$ $V_1 = wRs$ (α as for Case 1)	$H = \dfrac{wl^2}{8f}$ $V_1 = \dfrac{1}{2}wl$ $M_1 = 0$ (Ref. 1)

*From: Roark, *Formulas for Stress and Stress and Strain*, 4th ed., McGraw-Hill.
Reference 1: Leontovich, *Frames and Arches*, McGraw-Hill.

Table 5.1N continued on page 102

TABLE 5.3 Symmetrical Three-Hinged Arches of Any Shape: Formulas for Various Static Loading Conditions

Loadings	Support Reactions	Bending Moments
5.	$R_A = \dfrac{5}{24}wL$ $R_B = \dfrac{1}{24}wL$ $H_A = H_B = \dfrac{wL^2}{48f}$	$M_m = \dfrac{wL^2}{48}\left[2\xi_m + 8\left(\xi_{1m} - \xi_{1m}^3 - \xi_m + \xi_m^3\right) - \eta_m\right],$ $M_k = \dfrac{wL^2}{48}(2\xi_{1k} - \eta_k).$
6.	$R_A = R_B = \dfrac{wL}{4}$ $H_A = H_B = \dfrac{wL^2}{24f}$	$M_m = \dfrac{wL^2}{24}\left[2\xi_m + 4\left(\xi_{1m} - \xi_{1m}^3 - \xi_m + \xi_m^3\right) - \eta_m\right]$
7.	$R_A = -\dfrac{wf^2}{6L},\ R_B = \dfrac{wf^2}{6L}$ $H_A = -\dfrac{5}{12}wf,$ $H_B = \dfrac{1}{12}wf$	$M_m = \dfrac{wL^2}{12}\left[2\left(\xi_{1m} - \xi_{1m}^3\right) + \eta_m - 2\xi_m\right],$ $M_k = \dfrac{wL^2}{12}(2\xi_{1k} - \eta_k).$
8.	$R_A = R_B = 0$ $H_A = H_B = -\dfrac{M_0}{f}$	$M_m = M_0\eta_m$

─ **N O T E S** ─

Example for Table 5.4. Two-hinged parabolic arch

Given. Parabolic arch 3 in Table 5.4

$$L = 20 \text{ m}, f = 3 \text{ m}, x = a = 5 \text{ m}, \xi = \frac{a}{L} = \frac{5}{20} = 0.25$$

$$\tan \phi_x = \frac{4f(L-2x)}{L^2} = \frac{4 \times 3(20 - 2 \times 5)}{20^2} = 0.3$$

$$\phi_x = 16.7°, \sin \phi_x = 0.287, \cos \phi_x = 0.958$$

Concentrated load $P = 20$ kN

Required. Compute support reactions R_A and H_A, bending moments M_c and M_x, axial force N_x, and shear V_x (at point of load).

Solution. $R_A = P\frac{L-a}{L} = 20\frac{20-5}{20} = 15$ kN

$$H_A = \frac{5PL}{8f} k(\xi - 2\xi^2 + \xi^4) = \frac{5 \times 20 \times 20}{8 \times 3} \times 1(0.25 - 2 \times 0.25^2 + 0.25^4) = 10.75 \text{ kN}$$

$$M_c = \frac{PL}{8}[4\xi - 5k(\xi - 2\xi^3 + \xi^4)] = \frac{20 \times 20}{8}[4 \times 0.25 - 5(0.25 - 2 \times 0.25^3 + 0.25^4) = -9.5 \text{ kN} \cdot \text{m}]$$

$$y_x = \frac{4f(L-x)x}{L^2} = \frac{4 \times 3(20-5) \times 5}{20^2} = 2.25$$

$$M_x = R_A a - H_A y_x = 15 \times 5 - 10.75 \times 2.25 = 50.81 \text{ kN} \cdot \text{m}$$

$$N_x = R_A \sin \phi_x + H_A \cos \phi_x = 15 \times 0.287 + 10.75 \times 0.958 = 14.6 \text{ kN}$$

$$V_x = R_A \cos \phi_x - H_A \sin \phi_x = 15 \times 0.958 - 10.75 \times 0.287 = 11.3 \text{ kN}$$

TABLE 5.4 Two-Hinged Parabolic Arches: Formulas for Various Static Loading Conditions

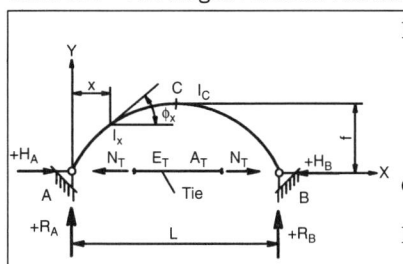

Equation of parabola:
$$y = \frac{4f(L-x)x}{L^2}, \quad I_x = I_c/\cos\phi_x$$
$$\tan\phi = \frac{dy}{dx} = \frac{4f(L-2x)}{L^2}$$

Coefficients: For regular arch: $\upsilon = 0$, $k = 1$

For tied arch: $\upsilon = \frac{15}{8} \cdot \frac{\beta}{f^2}$, $k = \frac{1}{1+\upsilon}$, $\beta = \frac{EI_c}{E_T A_T}$

Loadings	Support Reactions	Bending Moments
1. Uniformly distributed load w over full span	$R_A = R_B = \dfrac{wL}{2}$, $H_A = H_B = \dfrac{wL^2}{8f}k$.	$M_c = \dfrac{wL^2}{8}(1-k)$, $\upsilon = \dfrac{15}{8} \cdot \dfrac{\beta}{f^2}$, $k = \dfrac{1}{1+\upsilon}$.
2. Uniformly distributed load w over half span (at $L/4$, m)	$R_A = \dfrac{3}{8}wL$, $R_B = \dfrac{1}{8}wL$, $H_A = H_B = \dfrac{wL^2}{16f}k$.	$M_c = \dfrac{wL^2}{16}(1-k)$, $M_m = \left(\dfrac{1}{16} - \dfrac{3}{64}k\right)wL^2$.
3. Point load P at $a = \xi L$	$R_A = P\dfrac{L-a}{L}$, $R_B = P\dfrac{a}{L}$, $H_A = H_B = \dfrac{5PL}{8f}k(\xi - 2\xi^3 + \xi^4)$.	$M_c = \dfrac{PL}{8}[4\xi - 5k(\xi - 2\xi^3 + \xi^4)]$, $\xi = \dfrac{a}{L}$.
4. Triangular load w	$R_A = \dfrac{5wL}{24}$, $R_B = \dfrac{wL}{24}$, $H_A = H_B = 0.0228\dfrac{wL^2}{f}k$.	$M_C = R_B \dfrac{L}{2} - H_B f$.

TABLE 5.1N Formulas for Circular Rings and Arches* (Continued)

Loading, Support, and Case Number	Formulas for End Reactions	
	Circular Arch	Parabolic Arch; l = Span; f = Rise; a = Distance Load from Left End, b from Right
5. Ends pinned; concentrated load W at any point ϕ	$V = \frac{1}{2}W\left(\frac{n-s}{c}\right)$ $H_1 = W\left(\frac{\pi - \theta - \phi + \pi sn - 3sc - 2\theta s^2 + \pi s^2 - ne - 2se - 2ns}{\pi + 2\pi s^2 - 2\theta - 6sc - 4\theta s^2}\right)$	
6. Like Case 5 except ends fixed	$V = W\left(\frac{2\theta ec + \theta s^2 + \theta n^2 - 2\theta}{2\theta sc - 2\theta^2}\right)$ $H_2 = W\left(\frac{s^2 - \frac{1}{2}\theta en + \phi es - ns - \frac{1}{2}\theta sc - \frac{1}{2}\theta^2 + \frac{1}{2}\theta \phi}{2s^2 - \theta^2 - \theta sc}\right)$ $M_0 = WR\left(\frac{ec + \frac{1}{2}s^2 + \theta ne - \phi ne - sn - e^2 + \frac{1}{2}n^2}{2\theta n}\right) + \frac{H_2R(2sn - 2\theta en) + VR(\theta - sc + 2\theta n^2)}{2\theta n}$	
7. Ends pinned; uniform load w lb per linear in	$V = \frac{1}{4}wR\frac{(1-s)^2}{c}$ $H_1 = wR\left[\frac{\frac{3}{4}\theta s + \frac{2}{3}\frac{3}{2}s^2c - \frac{3}{8}\pi s + \frac{3}{4}\pi s^3 - \frac{1}{6}c + \theta s^3 + \frac{1}{2}\pi s c + s - 2c^2 - s^2 - 2cs - \frac{1}{2}}{\frac{3}{4}\pi + \frac{3}{2}\pi s^2 - \theta - \frac{3}{4}sc - 2s^2}\right]$ $M_0 = WR\left(\frac{ec + \frac{1}{2}s^2 + \theta ne - \phi ne - sn - e^2 + \frac{1}{2}n^2}{2\theta n}\right) + \frac{H_2R(2sn - 2\theta en) + VR(\theta - sc + 2\theta n^2)}{2\theta n}$	$V = \frac{1}{2}W\frac{f}{l}$ $H_1 = \frac{3}{1}W$ $(W = wf)$ **(Ref. 1)**
8. Like Case 7 except ends fixed	$V = wR\left(\frac{4 - 3c - 3s^2 + c^3}{6\theta - 6sc}\right)$ $H_2 = wR\left(\frac{\theta^2 - \frac{1}{6}\theta s + \theta sc - 2s^2 + \frac{1}{2}\theta sc^2}{2\theta^2 - 4s^2 + 2\theta sc}\right)$ $M_0 = \frac{1}{4}wR^2\left(\frac{2}{3} - 2\frac{s}{\theta} + \frac{1}{2}\frac{sc}{\theta}\right) - H_2R\left(1 - \frac{s}{\theta}\right)$	$V = \frac{1}{4}W\frac{f}{l}$ $H_2 = 2^{5 \cdot 5 \cdot 6}W$ $M_1 = 2^{30 \cdot 7 \cdot 4 \cdot 38}Wf$ $(W = wf)$ **(Ref. 1)**

Arches: Diagrams and Formulas for Various Loading Conditions

TABLE 5.5 Two-Hinged Parabolic Arches: Formulas For Various Static Loading Conditions

Loadings	Support Reactions	Bending Moments
5. (load W on left half, C at crown, supports A, B)	$R_A = -\dfrac{wf^2}{2L}$, $R_B = -R_A$ $H_A = -0.714wf$ $H_B = 0.286wf$	$M_C = -0.0357wf^2$
6. (triangular load W, peak at A)	$R_A = -\dfrac{wf^2}{6L}$, $R_B = -R_A$ $H_A = -0.401wf$ $H_B = 0.099wf$	$M_C = -0.0159wf^2$
7. Tied arch (load W on left half, Tie)	$R_A = -\dfrac{wf^2}{2L}$, $R_B = -R_A$ $H = wf$ $N_T = \dfrac{2.286wf^3}{8f^2 + 15\beta}$	$M_C = \dfrac{wf^2}{4} - N_T f$
8. Tied arch (triangular load W, Tie)	$R_A = -\dfrac{wf^2}{6L}$, $R_B = -R_A$ $H = \dfrac{wf}{2}$ $N_T = \dfrac{0.792wf^3}{8f^2 + 15\beta}$	$M_C = \dfrac{wf^2}{12} - N_T f$
9. (temperature/displacement effect)	$R_A = R_B = 0$ $H = \dfrac{15}{8} \cdot \dfrac{EI_C \Delta_L}{f^2 L} k$	$M_C = -Hf$

NOTES

Example for Table 5.6. Fixed parabolic arch

Given. Fixed parabolic arch 2 in Table 5.6

$$L = 20 \text{ m}, f = 3 \text{ m}, x = \xi L = 8 \text{ m}, \xi = \frac{8}{20} = 0.4, \xi_1 = \frac{L-x}{L} = \frac{20-8}{20} = 0.6$$

Distribution load $w = 2$ kN/m

Required. Compute support reactions R_A and H_A, bending moments M_A and M_C.

Solution. $R_A = \dfrac{wL}{2}\xi[1+\xi_1(1+\xi\xi_1)] = \dfrac{2\times 20}{2}0.4[1+0.6(1+0.4\times 0.6)] = 13.95$ kN

$$H_A = \frac{wL^2}{8f}\xi^3[1+3\xi_1(1+2\xi_1)]$$

$$= \frac{2\times 20^2}{8\times 3}\times 0.4^3 \times [1+2\times 0.6(1+2\times 0.6)] = 10.58 \text{ kN}$$

$$M_A = -\frac{wL^2}{2}\xi^2\xi_1^3 = -\frac{2\times 20^2}{2}\times 0.4^2 \times 0.6^3 = -13.82 \text{ kN}\cdot\text{m}$$

$$M_C = R_A \frac{L}{2} - w\times 8\times 6 - H_A f - M_A$$

$$= 13.95\times 10 - 2\times 8\times 6 - 10.58\times 3 - 13.82 = -2.06 \text{ kN}\cdot\text{m}$$

Arches: Diagrams and Formulas for Various Loading Conditions

TABLE 5.6 Fixed Parabolic Arches: Formulas for Various Static Loading Conditions

Equation of parabola:

$$y = \frac{4f(L-x)x}{L^2}, \quad I_x = I_c / \cos\phi_x$$

$$\tan\phi = \frac{dy}{dx} = \frac{4f(L-2x)}{L^2}$$

$$\xi = \frac{x}{L}, \quad \xi_1 = \frac{L-x}{L}$$

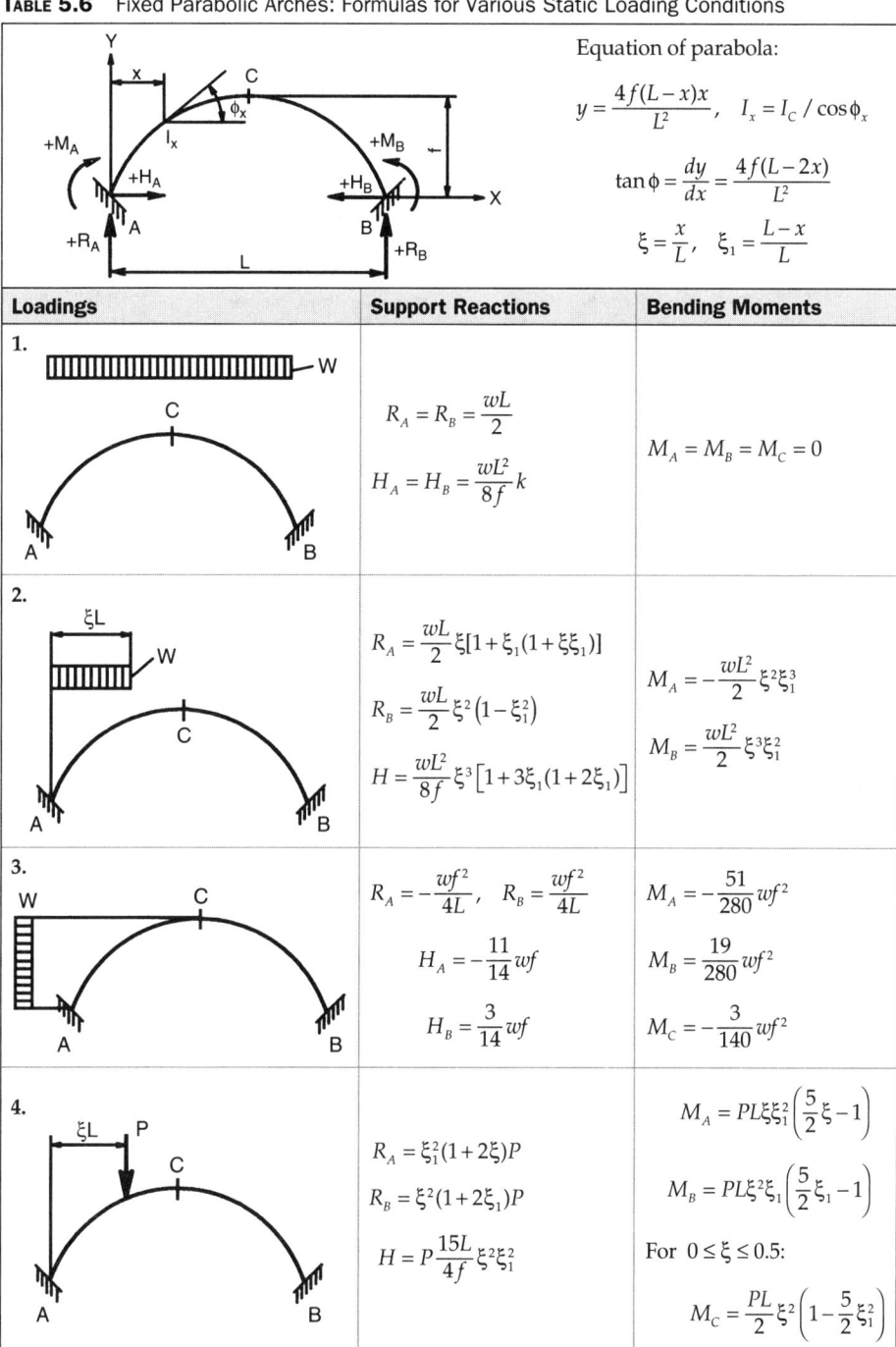

Loadings	Support Reactions	Bending Moments
1. (uniform load w over span)	$R_A = R_B = \dfrac{wL}{2}$ $H_A = H_B = \dfrac{wL^2}{8f} k$	$M_A = M_B = M_C = 0$
2. (partial uniform load w over ξL)	$R_A = \dfrac{wL}{2}\xi[1+\xi_1(1+\xi\xi_1)]$ $R_B = \dfrac{wL}{2}\xi^2(1-\xi_1^2)$ $H = \dfrac{wL^2}{8f}\xi^3\left[1+3\xi_1(1+2\xi_1)\right]$	$M_A = -\dfrac{wL^2}{2}\xi^2\xi_1^3$ $M_B = \dfrac{wL^2}{2}\xi^3\xi_1^2$
3. (triangular load W on left)	$R_A = -\dfrac{wf^2}{4L}, \quad R_B = \dfrac{wf^2}{4L}$ $H_A = -\dfrac{11}{14}wf$ $H_B = \dfrac{3}{14}wf$	$M_A = -\dfrac{51}{280}wf^2$ $M_B = \dfrac{19}{280}wf^2$ $M_C = -\dfrac{3}{140}wf^2$
4. (point load P at ξL)	$R_A = \xi_1^2(1+2\xi)P$ $R_B = \xi^2(1+2\xi_1)P$ $H = P\dfrac{15L}{4f}\xi^2\xi_1^2$	$M_A = PL\xi\xi_1^2\left(\dfrac{5}{2}\xi - 1\right)$ $M_B = PL\xi^2\xi_1\left(\dfrac{5}{2}\xi_1 - 1\right)$ For $0 \le \xi \le 0.5$: $M_C = \dfrac{PL}{2}\xi^2\left(1 - \dfrac{5}{2}\xi_1^2\right)$

―――――― N O T E S ――――――

Eccentrically Curved Beams

These beams (Fig. 5.1) are bounded by arcs having different centers of curvature. In addition, it is possible for either radius to be the larger one. The one in which the section depth shortens as the central section is approached may be called the *arch beam*. When the central section is the largest, the beam is of the crescent type.

Crescent I denotes the beam of larger outside radius and *Crescent II* of larger inside radius. The stress at the *central section* of such beams may be found from $S = KMCII$. In the case of rectangular cross section, the equation becomes $S = 6KM/bh^2$, where M is the bending moment, b is the width of the beam section, and h its height. The *stress factors* K for the *inner boundary*, established from photoelastic data, are given in Table 5.2N. The outside radius is denoted by R_o and the inside by R_i. The geometry of crescent beams is such that the stress can be larger in *off-center sections*. The stress at the central

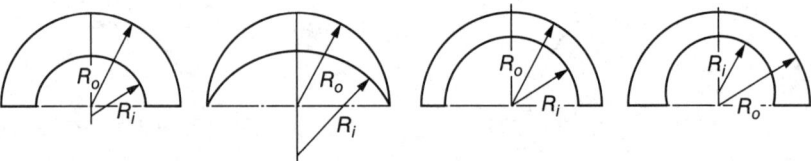

FIGURE 5.1 Eccentrically curved beams.

TABLE 5.2N Stress Factors for Inner Boundary at Central Section

1. For the arch-type beams
   (a) $K = 0.834 + 1.504 \dfrac{h}{R_o + R_i}$ if $\dfrac{R_o + R_i}{h} < 5$
   (b) $K = 0.899 + 1.181 \dfrac{h}{R_o + R_i}$ if $5 < \dfrac{R_o + R_i}{h} < 10$
   (c) In the case of larger section ratios, use the equivalent beam solution.

2. For the crescent I-type beams
   (a) $K = 0.570 + 1.536 \dfrac{h}{R_o + R_i}$ if $\dfrac{R_o + R_i}{h} < 2$
   (b) $K = 0.959 + 0.769 \dfrac{h}{R_o + R_i}$ if $2 < \dfrac{R_o + R_i}{h} < 20$
   (c) $K = 1.092 \left(\dfrac{h}{R_o + R_i} \right)^{0.0298}$ if $\dfrac{R_o + R_i}{h} > 20$

3. For the crescent II-type beams
   (a) $K = 0.897 + 1.098 \dfrac{h}{R_o + R_i}$ if $\dfrac{R_o + R_i}{h} < 8$
   (b) $K = 1.119 \left(\dfrac{h}{R_o + R_i} \right)^{0.0298}$ if $8 < \dfrac{R_o + R_i}{h} < 20$
   (c) $K = 1.081 \left(\dfrac{h}{R_o + R_i} \right)^{0.0270}$ if $\dfrac{R_o + R_i}{h} > 20$

Continued on page 108

Arches: Diagrams and Formulas for Various Loading Conditions

TABLE 5.7 Fixed Parabolic Arches: Formulas for Various Static Loading Conditions

Loadings	Support Reactions	Bending Moments
5. Point load P at crown C, with $L/2$ on each side; supports A, B	$R_A = R_B = \dfrac{P}{2}$ $H = \dfrac{15PL}{64f}$	$M_A = M_B = \dfrac{PL}{32}$ $M_C = \dfrac{3PL}{64}$
6. Triangular distributed load w peaking at supports, zero at crown	$R_A = R_B = \dfrac{wL}{4}$ $H = \dfrac{5wL^2}{128f}$	$M_A = M_B = -\dfrac{wL^2}{192}$ $M_C = -\dfrac{wL^2}{384}$
7. Rotation $\phi_A = 1$ at support A	$R_A = -\dfrac{6EI_C}{L^2}$ $R_B = +\dfrac{6EI_C}{L^2}$ $H = \dfrac{15}{2f}\cdot\dfrac{EI_C}{L}$	$M_A = \dfrac{9EI_C}{L}$ $M_B = \dfrac{3EI_C}{L}$ $M_C = -\dfrac{3}{2}\cdot\dfrac{EI_C}{L}$
8. Horizontal displacement = 1 at support A	$R_A = R_B = 0$ $H = \dfrac{45}{4}\cdot\dfrac{EI_C}{f^2 L}$	$M_A = M_B = \dfrac{15}{2f}\cdot\dfrac{EI_C}{L}$ $M_C = -\dfrac{15}{4f}\cdot\dfrac{EI_C}{L}$

NOTES

section determined above must then be multiplied by the *position factor k*, given in Table 5.3N. As in the concentric beam, the *neutral surface* shifts slightly toward the inner boundary. (See Vidosic, "Curved Beams with Eccentric Boundaries." *Transactions of the ASME*, 79, pp. 1317–1321.)

TABLE 5.3N Crescent-Beam Position Stress Factors*

	k	
Angle θ, deg	**Inner**	**Outer**
10	$1 + 0.055\, H/h$	$1 + 0.03\, H/h$
20	$1 + 0.164\, H/h$	$1 + 0.10\, H/h$
30	$1 + 0.365\, H/h$	$1 + 0.25\, H/h$
40	$1 + 0.567\, H/h$	$1 + 0.467\, H/h$
50	$1.521 - \dfrac{(0.5171 - 1.382\, H/h)^{1/2}}{1.382}$	$1 + 0.733\, H/h$
60	$1.756 - \dfrac{(0.2416 - 0.6506\, H/h)^{1/2}}{0.6506}$	$1 + 1.123\, H/h$
70	$2.070 - \dfrac{(0.4817 - 1.298\, H/h)^{1/2}}{0.6492}$	$1 + 1.70\, H/h$
80	$2.531 - \dfrac{(0.2939 - 0.7084\, H/h)^{1/2}}{0.3542}$	$1 + 2.383\, H/h$
90		$1 + 3.933\, H/h$

Note: All formulas are valid for $0 < H/h \leq 0.325$. Formulas for the inner boundary, except for 40 degrees, may be used to $H/h \leq 0.36$. H = distance between centers.

Arches: Diagrams and Formulas for Various Loading Conditions

TABLE 5.8 Three-Hinged Arches: Influence Lines

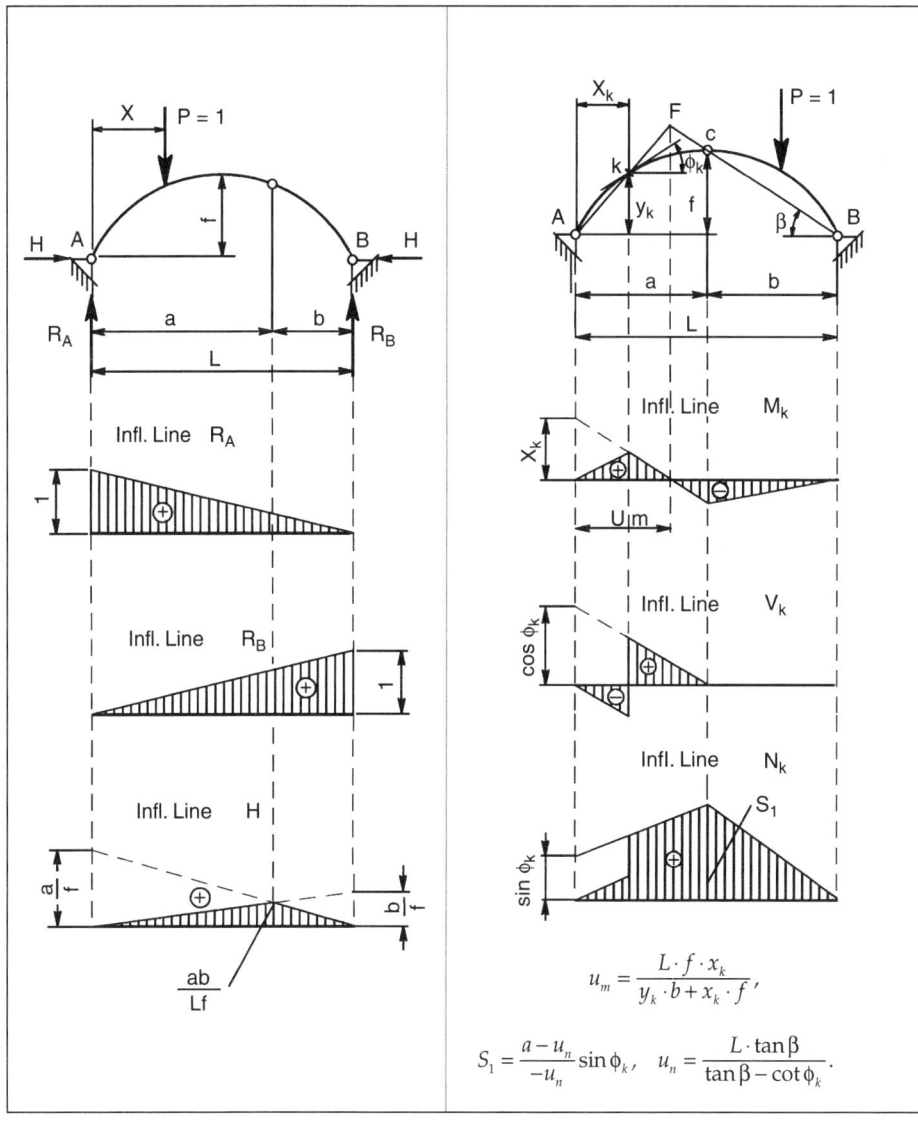

$$u_m = \frac{L \cdot f \cdot x_k}{y_k \cdot b + x_k \cdot f},$$

$$S_1 = \frac{a - u_n}{-u_n} \sin \phi_k, \quad u_n = \frac{L \cdot \tan \beta}{\tan \beta - \cot \phi_k}.$$

Example for Table 5.9. Fixed parabolic arch
Given. $L = 40$ m, $f = 10$ m, $x_k = 8$ m
Concentrated load in point k: $P_k = 12$ kN
Required. Using influence lines, compute support reactions R_A and H_A, support bending moment M_A, bending moments M_c and M_k, axial force N_k, and shear V_k.

Solution. $\dfrac{x_k}{L} = \dfrac{8}{40} = 0.2$, $y_k = \dfrac{4 \times 10(40-8)8}{40^2} = 6.4$ m

$\tan \phi = \dfrac{4f(L-2x)}{L^2} = \dfrac{4 \times 10(40 - 2 \times 8)}{40^2} = 0.6$

$\phi_k = 30.96°$, $\sin \phi_k = 0.514$, $\cos \phi_k = 0.857$

$R_A = S_i \times P_k = 0.896 \times 12 = 10.752$ kN

$H_A = S_i \times \dfrac{L}{f} \times P_k = 0.0960 \times \dfrac{40}{10} \times 12 = 4.608$ kN

$M_A = S_i \times L \times P_k = -0.0640 \times 40 \times 12 = 30.72$ kN·m

$M_c = S_i \times L \times P_k = -0.0120 \times 40 \times 12 = -5.76$ kN·m

$M_k = R_A \cdot x_k - H_A \cdot y_k - M_A = 10.752 \times 8 - 4.608 \times 6.4 - 30.72 = 25.805$ kN·m

$N_k = R_A \sin \phi_k + H_A \cos \phi_k = 10.752 \times 0.514 + 4.608 \times 0.857 = 9.475$ kN

$V_k = R_A \cos \phi_k - H_A \sin \phi_k = 10.752 \times 0.857 - 4.608 \times 0.514 = 6.745$ kN

Arches: Diagrams and Formulas for Various Loading Conditions

TABLE 5.9 Fixed Parabolic Arches: Influence Lines

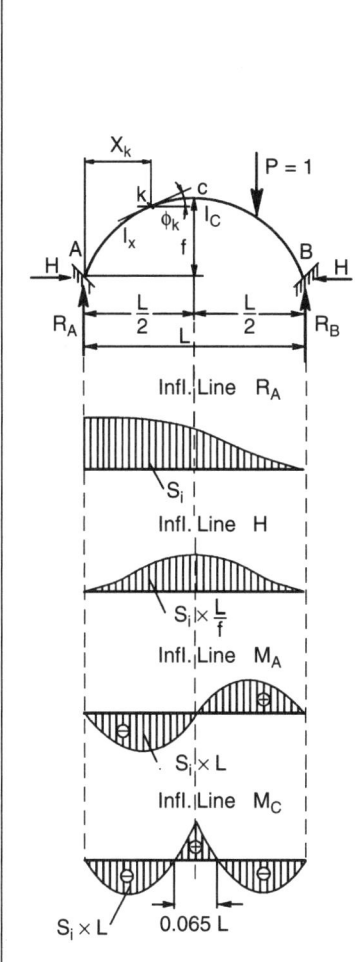

Equation of parabola: $y = \dfrac{4f(L-x)x}{L^2}$

$I_x = I_C/\cos\phi_x$, $\tan\phi = \dfrac{dx}{dy} = \dfrac{4f(L-2x)}{L^2}$

$R_A = S_i \times P$, $H = S_i \times \dfrac{L}{f} \times P$, $M = S_i \times L \times P$

Ordinates of Influence Lines (S_i)

$\dfrac{x}{L}$	R_A	H	M_A	M_C
0.0	1.000	0.0	0.0	0.0
0.05	0.993	0.0085	−0.0395	−0.0016
0.10	0.972	0.0305	−0.0625	−0.0052
0.15	0.939	0.0610	−0.0678	−0.0090
0.20	0.896	0.0960	−0.0640	−0.0120
0.25	0.844	0.1320	−0.0528	−0.0127
0.30	0.784	0.1655	−0.0368	−0.0102
0.35	0.718	0.1940	−0.0184	−0.0034
0.40	0.648	0.2160	0.0	0.0080
0.45	0.575	0.2295	0.0174	0.0246
0.50	0.500	0.2344	0.0312	0.0468
0.55	0.425	0.2295	0.0418	0.0246
0.60	0.352	0.2160	0.0480	0.0080
0.65	0.282	0.1940	0.0498	−0.0034
0.70	0.216	0.1655	0.0473	−0.0102
0.75	0.156	0.1320	0.0410	−0.0127
0.80	0.104	0.0960	0.0320	−0.0120
0.85	0.061	0.0610	0.0215	−0.0090
0.90	0.028	0.0305	0.0118	−0.0052
0.95	0.007	0.0085	0.0032	−0.0016
1.00	0.0	0.0	0.0	0.0

Reactions of a Three-Hinged Arch

The parabolic arch in Fig. 5.2 is hinged at A, B, and C. Determine the magnitude and direction of the reactions at the supports.

Calculation Procedure

1. *Consider the entire arch as a free body and take moments.*
   Since a moment cannot be transmitted across a hinge, the bending moments at A, B, and C are zero. Resolve the reactions R_A and R_C (Fig. 5.2) into their horizontal and vertical components.
   Considering the entire arch ABC as a free body, take moment with respect to A and C. Thus $\Sigma M_A = 8(10) + 10(25) + 12(40) + 8(56) - 5(25.2) - 72R_{CV} - 10.8R_{CH} = 0$, or $72R_{CV} + 10.8R_{CH} = 1132$, Eq. (a). Also, $\Sigma M_C = 72R_{AV} - 10.8R_{AH} - 8(62) - 10(47) - 12(32) - 8(16) - 5(14.4) = 0$, or $72R_{AV} - 10.8R_{AH} = 1550$, Eq. (b).

2. *Consider a segment of the arch and take moments.*
   Considering the segment BC as a free body, take moments with respect to B. Then $\Sigma M_B = 8(16) + 5(4.8) - 32R_{CV} + 19.2R_{CH} = 0$, or $32R_{CV} - 19.2R_{CH} = 152$, Eq. (c).

3. *Consider another segment and take moments.*
   Considering segment AB as a free body, take moments with respect to B: $\Sigma M_B = 40R_{AV} - 30R_{AH} - 8(30) - 10(15) = 0$, or $40R_{AV} - 30R_{AH} = 300$, Eq. (d).

4. *Solve the simultaneous moment equations.*
   Solve Eqs. (b) and (d) to determine R_A; solve Eqs. (a) and (c) to determine R_C. Thus $R_{AV} = 24.4$ kips (108.5 kN); $R_{AH} = 19.6$ kips (87.2 kN); $R_{CV} = 13.6$ kips (60.5 kN); $R_{CH} = 14.6$ kips (64.9 kN). Then $R_A = [(24.4)^2 + (19.6)^2]^{0.5} = 31.3$ kips (139.2 kN). Also $R_C = [(13.6)^2 + (14.6)^2]0.5 = 20.0$ kips (8.90 kN). And $\theta_A = $ arctan $(24.4/19.6) = 51°14'$; $\theta_C = $ arctan $(13.6/14.6)$ $42°58'$.

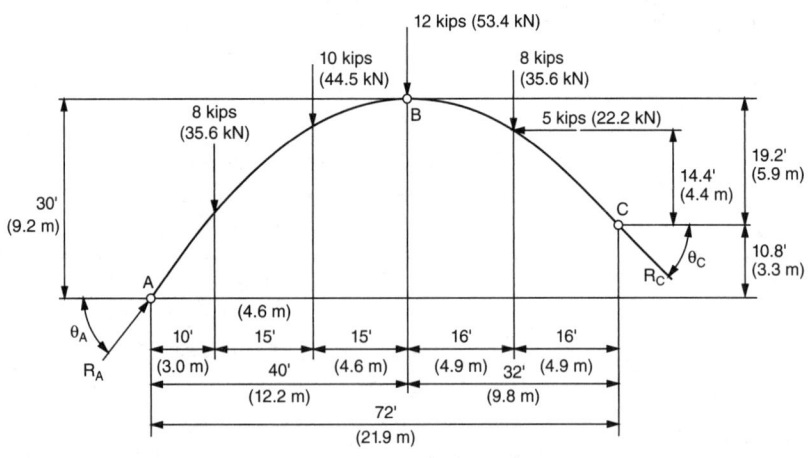

FIGURE 5.2 Parabolic arch.

Arches: Diagrams and Formulas for Various Loading Conditions

TABLE 5.10 Steel Rope

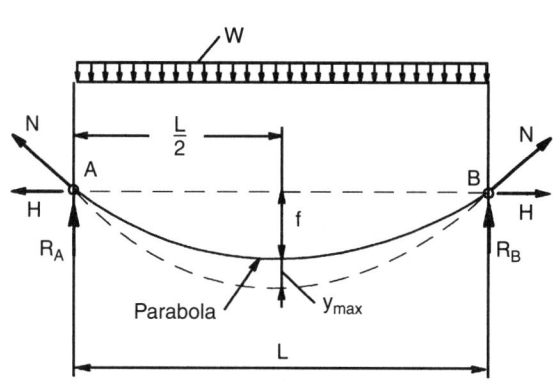

Rope Deflection

w = uniformly distributed load, f = rope sag due to natural weight, ($f \approx 1/20 \cdot L$)

s = length of rope, $s = \sqrt{L^2 + \dfrac{16}{3} f^2}$

Forces and Deflection

$H = \dfrac{\sqrt{0.25 wL^4}}{4f}$ (elastic deformations are not included)

$H = \sqrt[3]{\dfrac{w^2 L^2 EA}{24}}$ (elastic deformations are included)

E = modulus of elasticity, A = area of rope cross section

$N_{max} = \sqrt{H^2 + R^2}$, R = reaction, $R = wL/2$

Bending moment $M_{max} = wL^2/8$, deflection $y_{max} = \dfrac{M_{max}}{H}$

Temperature

$N_t = \alpha \cdot \Delta t^0 \cdot EA$, $\Delta t^0 = T_1^0 - T_2^0$, if: $\Delta t^0 > 0$ (tension), $\Delta t^0 < 0$ (compression)

α = linear coefficient of expansion

$H_t^3 - N_t \cdot H_t^2 = \dfrac{wL^2 EA}{24}$ $H_t^3 - N_t H_t^2 = \dfrac{wL^2 EA}{24}$, $N_{max} = \sqrt{H_t^2 + R^2}$, $y_{max} = \dfrac{M_{max}}{H_t}$

Length of Cable Carrying Known Loads

A cable is supported at points P and Q (Fig. 5.3a) and carries two vertical loads, as shown. If the tension in the cable is restricted to 1800 lb (8006 N), determine the minimum length of cable required to carry the loads.

Calculation Procedure:

1. *Sketch the loaded cable.*

   Assume a position of the cable, such as $PRSQ$ (Fig. 5.3a). In Fig. 5.3b, locate points P' and Q', corresponding to P and Q, respectively, in Fig. 5.3a.

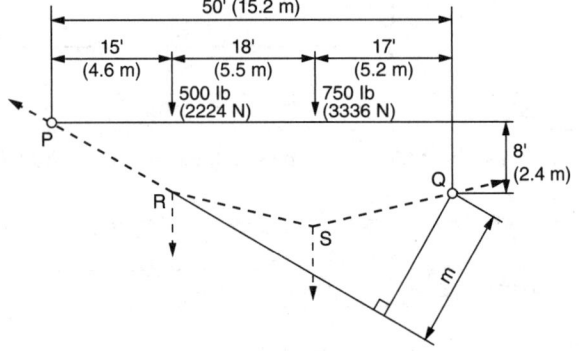

(a) Assumed position of loaded cable

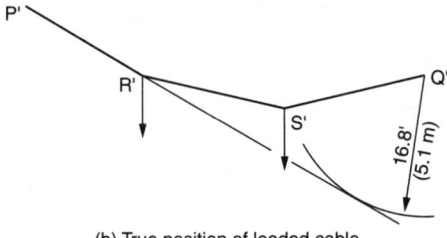

(b) True position of loaded cable

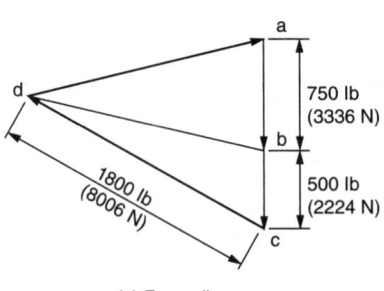

(c) Force diagram

Figure 5.3 Loaded cable analysis.

─────────────── N O T E S ───────────────

2. *Take moments with respect to an assumed point.*
   Assume that the maximum tension of 1800 lb (8006 N) occurs in segment *PR* (Fig. 5.3). The reaction at *P*, which is collinear with *PR*, is therefore 1800 lb (8006 N). Compute the true perpendicular distance *m* from *Q* to *PR* by taking moments with respect to *Q*. Or $\Sigma M_Q = 1800m - 500(35) - 750(17) = 0$; $m = 16.8$ ft (5.1 m). This dimension establishes the true position of *PR*.

3. *Start the graphical solution of the problem.*
   In Fig. 5.3b draw a circular arc having *Q'* as center and a radius of 16.8 ft (5.1 m). Draw a line through *P'* tangent to this arc. Locate *R'* on this tangent at a horizontal distance of 15 ft (4.6 m) from *P'*.

4. *Draw the force vectors.*
   In Fig. 5.3c draw vectors **ab**, **bc**, and **cd** to represent the 750-lb (3336-N) load, the 500-lb (2224-N) load, and the 1800-lb (8006-N) reaction at *P*, respectively. Complete the triangle by drawing vector **da**, which represents the reaction at *Q*.

5. *Check the tension assumption.*
   Scale *da* to ascertain whether it is less than 1800 lb (8006 N). This is found to be so, and the assumption that the maximum tension exists in *PR* is validated.

6. *Continue this construction.*
   Draw a line through *Q'* in Fig. 5.3b paraliel to *da* in Fig. 5.3c. Locate *S'* on this line at a horizontal distance of 17 ft (5.2 m) from *Q*.

7. *Complete the construction.*
   Draw *R'S'* and *db*. Test the accuracy of the construction by determining whether these lines are parallel.

8. *Determine the required length of the cable.*
   Obtain the required length of the cable by scaling the lengths of the segments to Fig. 5.3b. Thus $P'R' = 17.1$ ft (5.2 m); $R'S' = 18.4$ ft (5.6 m); $S'Q' = 17.6$ ft (5.4 m); and length of cable = 53.1 ft (16.2 m).

CHAPTER 6
Trusses: Method of Joints and Method of Section Analysis

NOTES

Tables 6.1 to 6.4 provide examples of analysis of flat trusses.

Legend Upper chord: U
        Lower chord: L
        Vertical posts: $U_i\text{-}L_i$
        Diagonals: $U_i\text{-}L_{i\pm 1}$
        End posts: $L_0\text{-}U_1$
        Load on upper chord: P^t
        Load on lower chord: P^b

The method of joints and the method of section analysis are used to compute forces in truss elements without relying on the computer. The method of joints is based on the equilibrium of the forces acting within the joint. The method of section analysis is based on the equilibrium of the forces acting from either the left or the right of the section $\left(\sum x = 0, \sum y = 0, \sum M = 0\right)$.

The truss joints are assumed to be hinges, and the loads acting on the truss are repressented as forces concentrated within the truss joints.

TABLE 6.1 Trusses: Method of Joints and Method of Section Analysis

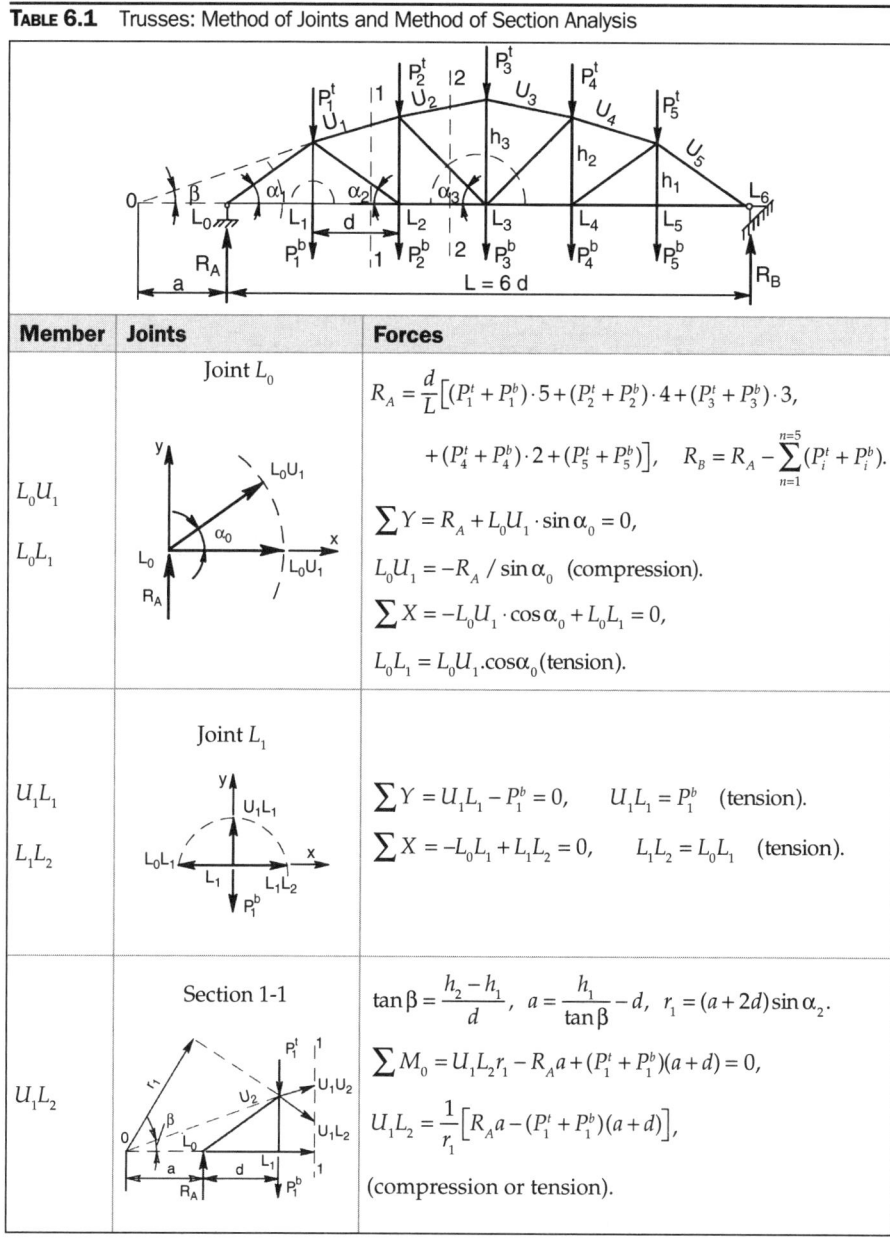

Member	Joints	Forces
L_0U_1 L_0L_1	Joint L_0	$R_A = \dfrac{d}{L}\left[(P_1^t + P_1^b)\cdot 5 + (P_2^t + P_2^b)\cdot 4 + (P_3^t + P_3^b)\cdot 3,\right.$ $\left. + (P_4^t + P_4^b)\cdot 2 + (P_5^t + P_5^b)\right], \quad R_B = R_A - \sum_{n=1}^{n=5}(P_i^t + P_i^b).$ $\sum Y = R_A + L_0U_1 \cdot \sin\alpha_0 = 0,$ $L_0U_1 = -R_A / \sin\alpha_0 \quad \text{(compression)}.$ $\sum X = -L_0U_1 \cdot \cos\alpha_0 + L_0L_1 = 0,$ $L_0L_1 = L_0U_1 \cdot \cos\alpha_0 \text{ (tension)}.$
U_1L_1 L_1L_2	Joint L_1	$\sum Y = U_1L_1 - P_1^b = 0, \quad U_1L_1 = P_1^b \quad \text{(tension)}.$ $\sum X = -L_0L_1 + L_1L_2 = 0, \quad L_1L_2 = L_0L_1 \quad \text{(tension)}.$
U_1L_2	Section 1-1	$\tan\beta = \dfrac{h_2 - h_1}{d}, \quad a = \dfrac{h_1}{\tan\beta} - d, \quad r_1 = (a + 2d)\sin\alpha_2.$ $\sum M_0 = U_1L_2 r_1 - R_A a + (P_1^t + P_1^b)(a + d) = 0,$ $U_1L_2 = \dfrac{1}{r_1}\left[R_A a - (P_1^t + P_1^b)(a + d)\right],$ (compression or tension).

Common Types of Trusses

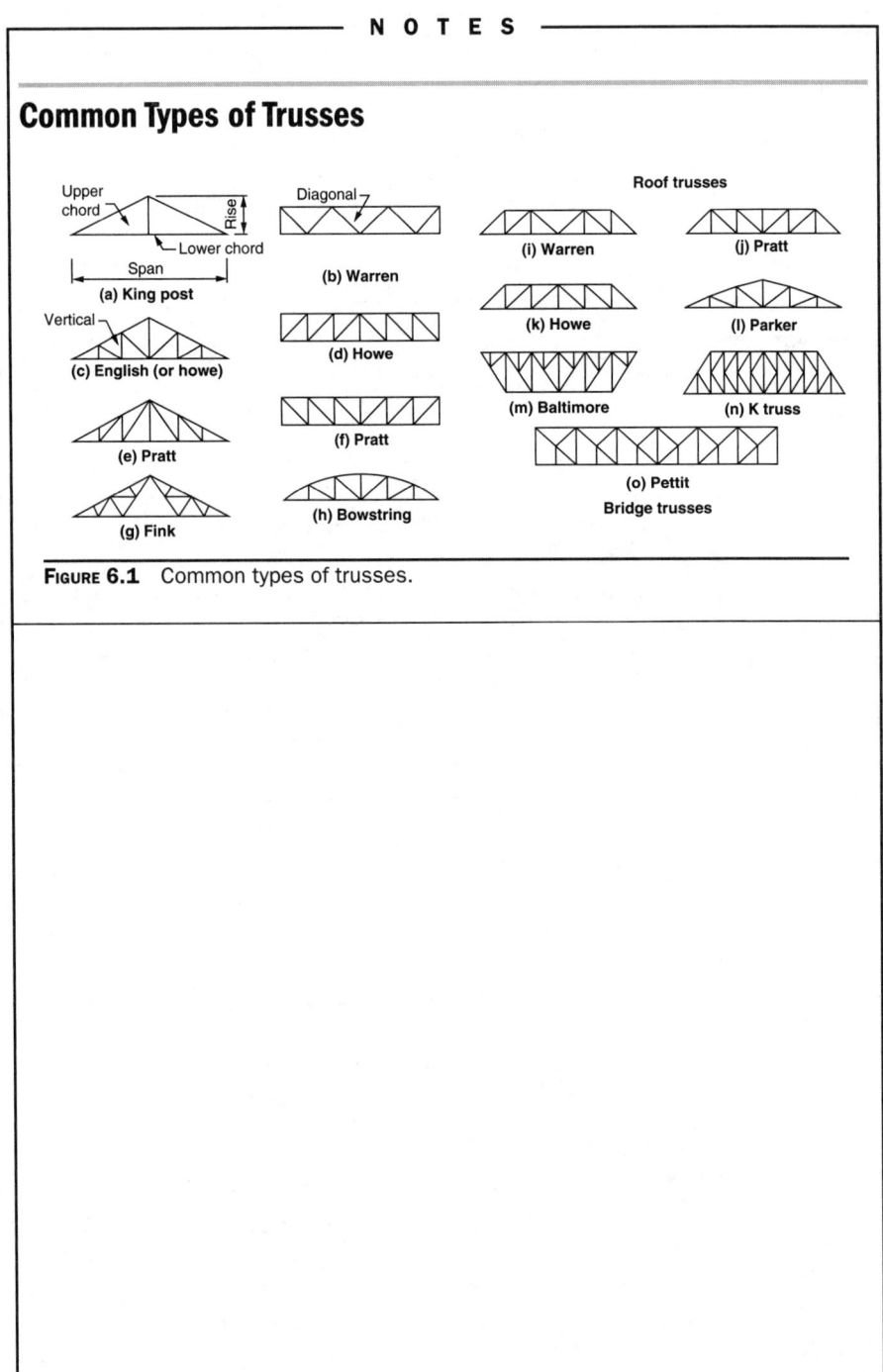

FIGURE 6.1 Common types of trusses.

TABLE 6.2 Trusses: Method of Joints and Method of Section Analysis

Member	Joints	Forces
U_1U_2	Section 1-1	$r_2 = (a+2d)\sin\beta$, $\sum M_{L2} = U_1U_2 r_2 + R_A 2d - (P_1^t + P_1^b)d = 0$, $U_1U_2 = -R_A 2d - (P_1^t + P_1^b)d$ (compression).
U_2L_2 L_2L_3	Joints L_2	$\sum Y = U_2L_2 - U_1L_2 \sin\alpha_2 - P_2^b = 0$, $U_2L_2 = P_2^b + U_1L_2 \sin\alpha_2$ (tension). $\sum X = -L_1L_2 + L_2L_3 + U_1L_2 \cos\alpha_2 = 0$, $L_2L_3 = L_1L_2 - U_1L_2 \cos\alpha_2$ (tension).
U_2L_3 L_2L_3	Section 2-2	$r_3 = (a+3d)\sin\alpha_3$ $\sum M_0 = U_2L_3 r_3 - R_A a + (P_1^t + P_b^1)(a+d)$ $+ (P_2^t + P_2^b)(a+2d) = 0$, (compression), $U_2L_3 = \dfrac{1}{r_3}\left[R_A a - (P_1^t + P_1^b)(a+d) - (P_2^t + P_2^b)(a+2d)\right]$, $\sum M_{U2} = -L_2L_3 h_2 + R_A 2d - (P_1^t + P_1^b)d = 0$, $L_2L_3 = \dfrac{1}{h_2}\left[R_A 2d - (P_1^t + P_1^b)d\right]$ (tension).
U_3L_3	Joint L_3	If $P_4^t = P_2^t$, $P_5^t = P_1^t$, $P_4^b = P_2^b$, $P_5^b = P_1^b$, $L_3L_4 = L_2L_3$, $U_4L_3 = U_2L_3$, $\sum Y = U_3L_3 - U_2L_3 \sin\alpha_3 - U_4L_3 \sin\alpha_3 - P_3^b = 0$, $U_3L_3 = P_3^b + U_2L_3 \sin\alpha_3 + U_4L_3 \sin\alpha_3$ (tension).

─────── N O T E S ───────

Determining Stress in a Truss Diagonal

Stress in a truss diagonal is determined by taking a vertical section and computing moments about the intersection of top and bottom chords.

Generally, the calculation can be simplified by determining first the vertical component of the diagonal and from it the stress. So resolve Bc into its horizontal and vertical components Bc_H and Bc_V, at c, so that the line of action of the horizontal component passes through 0. Taking moments about 0 yields

$$(Bc_V \times 0c) - (R \times 0a) + (P_1 \times 0b) = 0 \quad (6.1)$$

from which Bc_V may be determined. The actual stress in Bc is Bc_V multiplied by the secant of the angle that Bc makes with the vertical.

The stress in verticals, such as Cc, can be found in a similar manner. But take the section on a slope so as not to cut the diagonal but only the vertical and the chords. The moment equation about the intersection of the chords yields the stress in the vertical directly since it has no horizontal component. **Figures 6.2 and 6.3 show truss analysis methods.**

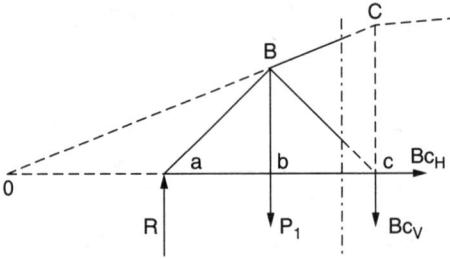

Figure 6.2 Truss diagonal.

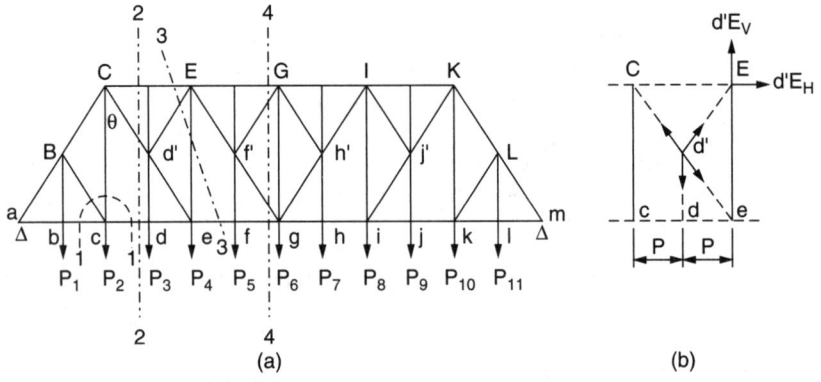

Figure 6.3 Sections taken through truss with subdivided panels for finding stresses in web members.

TABLE 6.3 Trusses: Influence Lines (Examples)

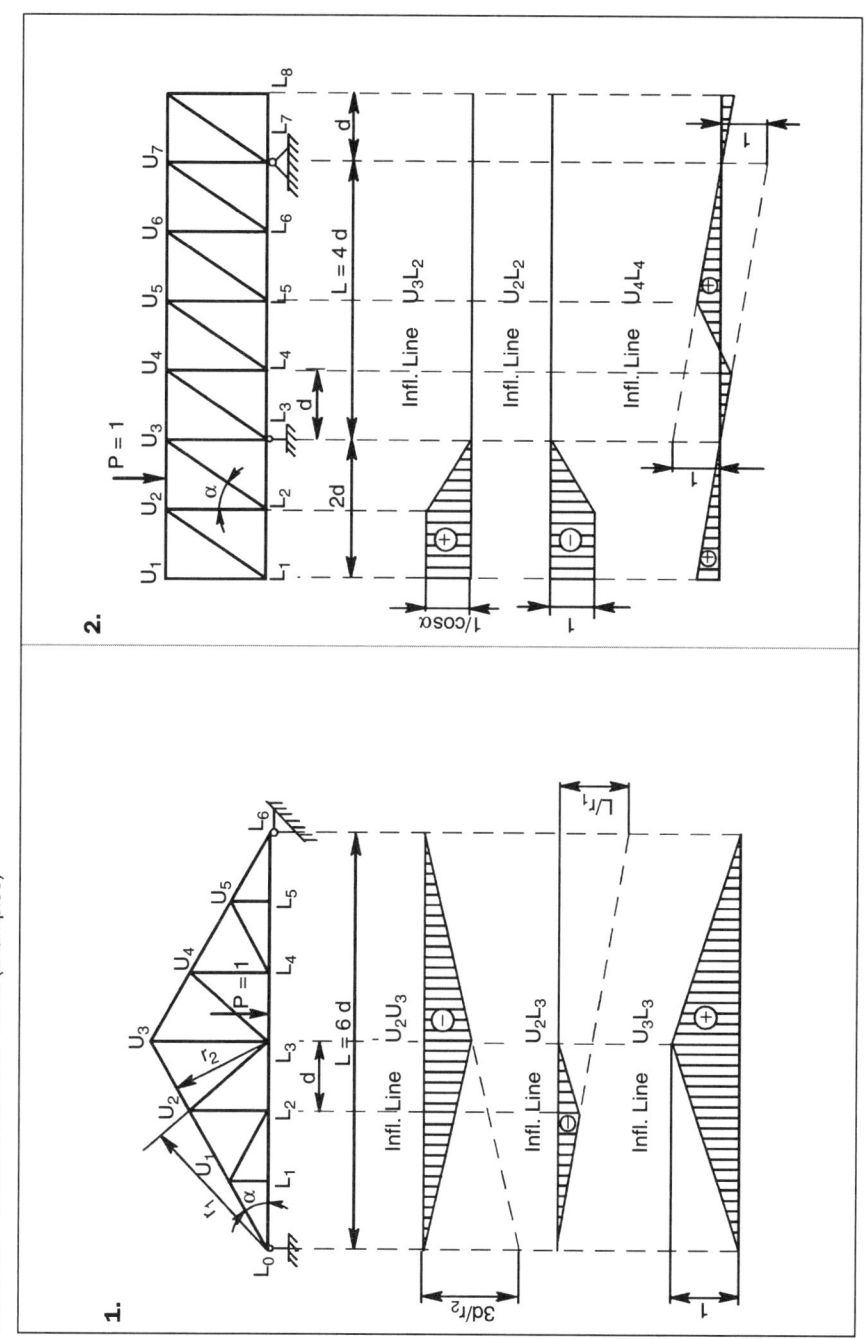

─── NOTES ───

Example for Table 6.4. Computation of truss

Given. Truss 3 in Table 6.4, $L = 12$ m, $d = 2$ m, $h = 4$ m

$$\tan\alpha = \frac{h}{d} = 2.0, \quad \alpha = 63.435°, \quad \cos\alpha = 0.447$$

Load: $P_2^b = P_6^b = 3$ kN, $P_3^b = P_5^b = 4$ kN, $P_4^b = 5$ kN

Required. Compute forces in truss members, using influence lines.

Solution.
$$\frac{2d}{h} = \frac{2\times 2}{4} = 1, \quad \frac{d}{h} = \frac{2}{4} = 0.5, \quad \frac{1}{\cos\alpha} = 2.237$$

$$U_4 U_5 = \frac{2d}{h} P_4^b + 2\frac{2d}{h(0.5L)} \times 2d \times P_3^b + 2\frac{2d}{h(0.5L)} \times d \times P_2^b$$

$$= 5 + 1.333 \times 4 + 0.667 \times 3 = 12.33 \text{ kN (compression)}$$

$$L_2 L_3 = \frac{d}{hL} \times d\ (5\times P_2^b + 4\times P_3^b + 3\times P_4^b + 2\times P_5^b + P_6^b)$$

$$= 0.083\ (5\times 3 + 4\times 4 + 3\times 5 + 2\times 4 + 3) = 4.73 \text{ kN (tension)}$$

$$U_2 L_3 = \frac{2.237}{L} d\ (-P_2^b + P_6^b + 2\times P_5^b + 3\times P_4^b + 4\times P_3^b)$$

$$= 0.3728\ (-3 + 3 + 2\times 4 + 3\times 5 + 4\times 4) = 14.53 \text{ kN (tension)}$$

$$U_4 L_3 = -U_2 L_3 = -14.57 \text{ kN \quad (compression)}$$

TABLE 6.4 Trusses: Influence Lines (Examples)

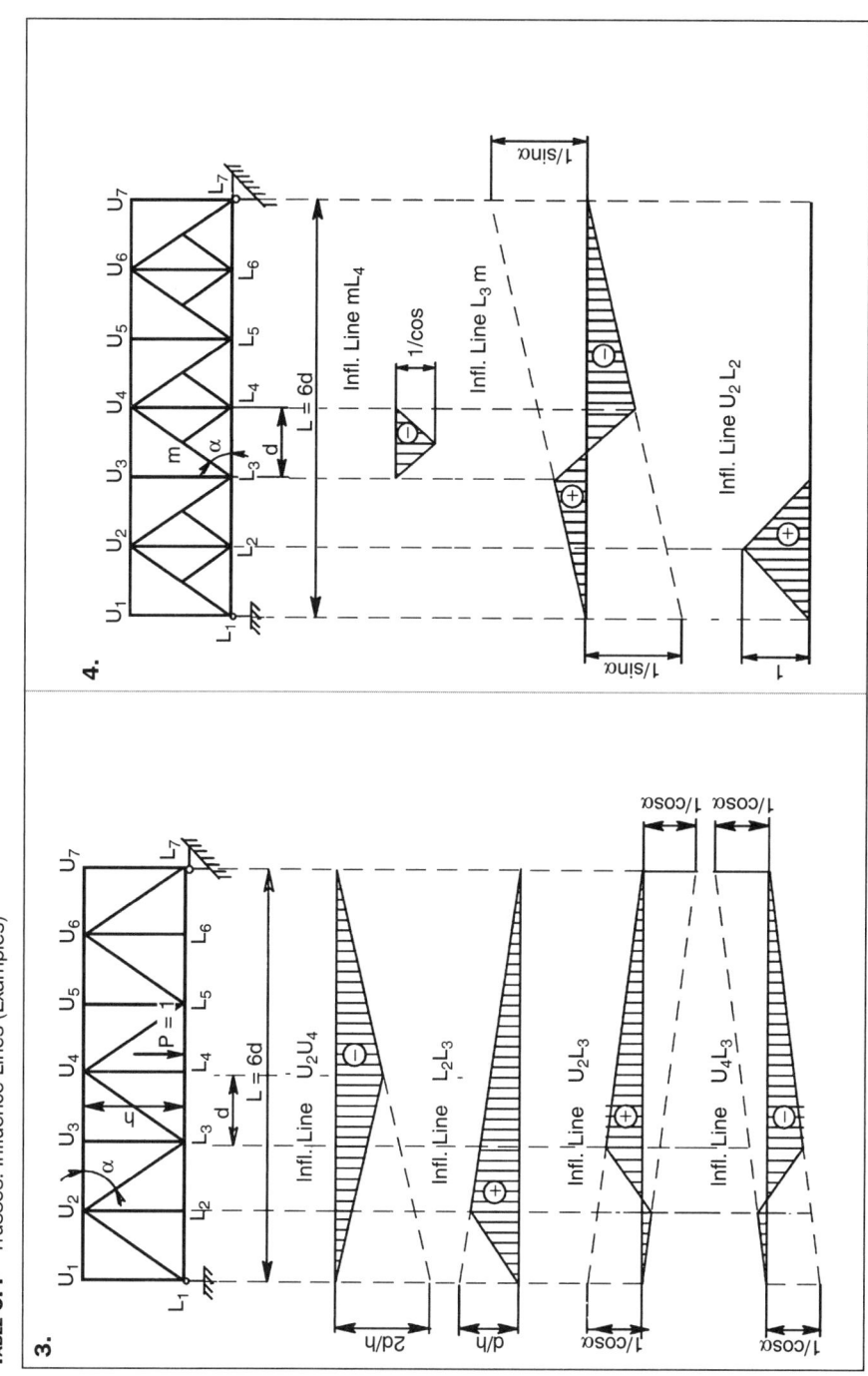

─────── N O T E S ───────

Truss Stresses Produced by Moving Loads

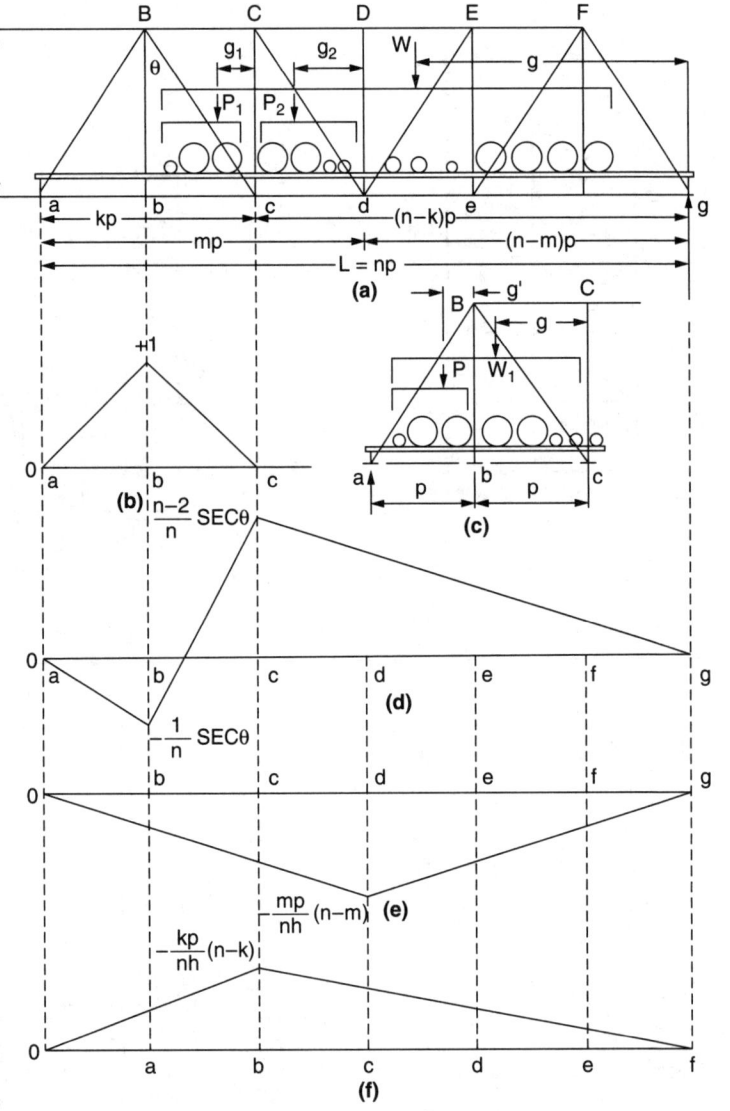

Figure 6.4 Stresses produced in a truss by moving loads are determined with influence lines.

CHAPTER 7
Plates: Bending Moments for Various Support and Loading Conditions

Statics

NOTES

Tables 7.1 to 7.9 provide formulas and coefficients for computation of bending moments in elastic plates.

The calculations are performed for plates of 1-m width.

The plates are analyzed in two directions for various support conditions and acting loads.

Units of measurement: Distributed loads (w): kN/m^2
                                      Bending moments (M): kN · m/m

Plates: Bending Moments for Various Support and Loading Conditions

TABLE 7.1 Rectangular Plates: Bending Moments

CASE A: $\dfrac{b}{a} > 2$ CASE B: $\dfrac{b}{a} \leq 2$

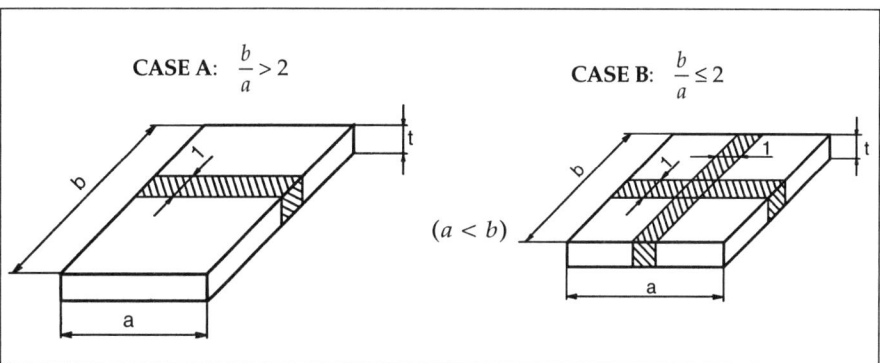

$(a < b)$

Case A $\dfrac{b}{a} > 2$ Plate should be computed in one (short) direction as a beam of length $L = a$.

Case B $\dfrac{b}{a} \leq 2$ Plate should be computed in two directions as two beams of lengths
$L_1 = a$ and $L_2 = b$.

Formulas for Bending Moment Computation $\left(\dfrac{b}{a} \leq 2\right)$

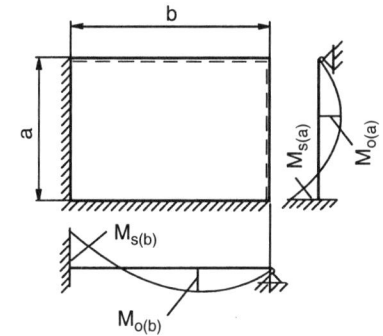

$M_{0(a)} = \alpha_a \cdot w \cdot a \cdot b, \quad M_{0(b)} = \alpha_b \cdot w \cdot a \cdot b$

$M_{s(a)} = \beta_a \cdot w \cdot a \cdot b, \quad M_{s(b)} = \beta_b \cdot w \cdot a \cdot b$

where w = uniformly distributed load
$\alpha_a, \alpha_b, \beta_a, \beta_b$ = coefficients from tables for
Poisson's ratio $\mu_T = 0$

Bending moments for any Poisson's ratio μ:

$M^\mu_{(a)} = \dfrac{1}{1-\mu_T^2}[(1-\mu\mu_T)M_{(a)} + (\mu-\mu_T)M_{(b)}], \quad M^\mu_{(b)} = \dfrac{1}{1-\mu_T^2}[(1-\mu\mu_T)M_{(b)} + (\mu-\mu_T)M_{(a)}]$

Legend:

Support condition
////////// — Plate fixed along edge.
– – – – – Plate hinged along edge.
——— Plate free along edge.
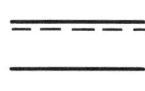 Plate supported on column.

NOTES

Example. Computation of rectangular plate, $b \leq 2a$

Given. Elastic steel plate 3 in Table 7.2, $a = 1.5$ m, $b = 2.1$ m, $t = 0.04$ m, $b/a = 1.4$
Uniformly distributed load $w = 0.8$ kN/m^2
Poisson's ratio $\mu = \mu_T = 0$

Required. Compute bending moments $M_{0(a)}$, $M_{0(b)}$, $M_{s(a)}$, $M_{s(b)}$.

Solution. $M_{0(a)} = \alpha_a wab = 0.0323 \times 0.8 \times 1.5 \times 2.1 = 0.0814$ kN·m/m = 81.4 N·m/m

$M_{0(b)} = \alpha_b wab = 0.0165 \times 0.8 \times 1.5 \times 2.1 = 0.0416$ kN·m/m = 41.6 N·m/m

$M_{s(a)} = \beta_a wab = -0.0709 \times 0.8 \times 1.5 \times 2.1 = -0.1787$ kN·m/m = −178.7 N·m/m

$M_{s(b)} = \beta_b wab = -0.0361 \times 0.8 \times 1.5 \times 2.1 = -0.0910$ kN·m/m = −91.0 N·m/m

Plates: Bending Moments for Various Support and Loading Conditions

TABLE 7.2 Rectangular Plates: Bending Moments (Uniformly Distributed Load)

Plate Supports	b/a	α_a	α_b	β_a	β_b
1.	1.0	0.0363	0.0365		
	1.1	0.0399	0.0330		
	1.2	0.0428	0.0298		
	1.3	0.0452	0.0268		
	1.4	0.0469	0.0240		
	1.5	0.0480	0.0214		
	1.6	0.0485	0.0189		
	1.7	0.0488	0.0169		
	1.8	0.0485	0.0148		
	1.9	0.0480	0.0133		
	2.0	0.0473	0.0118		
2.	1.0	0.0267	0.0180	−0.0694	
	1.1	0.0266	0.0146	−0.0667	
	1.2	0.0261	0.0118	−0.0633	
	1.3	0.0254	0.0097	−0.0599	
	1.4	0.0245	0.0080	−0.0565	
	1.5	0.0235	0.0066	−0.0534	
	1.6	0.0226	0.0056	−0.0506	
	1.7	0.0217	0.0047	−0.0476	
	1.8	0.0208	0.0040	−0.0454	
	1.9	0.0199	0.0034	−0.0432	
	2.0	0.0193	0.0030	−0.0412	
3.	1.0	0.0269	0.0269	−0.0625	−0.0625
	1.1	0.0292	0.0242	−0.0675	−0.0558
	1.2	0.0309	0.0214	−0.0703	−0.0488
	1.3	0.0319	0.0188	−0.0711	−0.0421
	1.4	0.0323	0.0165	−0.0709	−0.0361
	1.5	0.0324	0.0144	−0.0695	−0.0310
	1.6	0.0321	0.0125	−0.0678	−0.0265
	1.7	0.0316	0.0109	−0.0657	−0.0228
	1.8	0.0308	0.0096	−0.0635	-0.0196
	1.9	0.0302	0.0084	−0.0612	−0.0169
	2.0	0.0294	0.0074	−0.0588	−0.0147

Column Base Plates

AISC ASD Approach

The lowest columns of a structure usually are supported on a concrete foundation. To prevent crushing of the concrete, baseplates are inserted between the steel and concrete to distribute the load. For very heavy loads, a grillage, often encased in concrete, may be required. It consists of one or more layers of steel beams with pipe separators between them and tie rods through the pipe to prevent separation.

The area (in^2) of baseplate required may be computed from

$$A = \frac{P}{F_p} \tag{7.1}$$

where P = load, kips, and
F_p = allowable bearing pressure on support, ksi.

The allowable pressure depends on the strength of the concrete in the foundation and relative sizes of baseplate and concrete support area. If the baseplate occupies the full area of the support, then $F_p = 0.35 f'_c$ where f'_c is the 28-day compressive strength of the concrete. If the baseplate covers less than the full area, $F_p = 0.35 f'_c \sqrt{A_2/A_1} \leq 0.70 f'_c$, where A_1 is the baseplate area ($B \times N$) and A_2 is the full area of the concrete support.

Eccentricity of loading or presence of bending moment of the column base increases the pressure on some parts of the baseplate and decreases it on other parts. To compute these effects, the baseplate may be assumed completely rigid so that the pressure variation on the concrete is linear.

Plate thickness may be determined by treating projections m and n of the baseplate beyond the column as cantilevers. The cantilever dimensions m and n are usually defined as shows in Fig. 7.1. (If the baseplate is small, the area of the baseplate inside the column profile should be treated as a beam.) Yield-line analysis shows that an equivalent cantilever dimension n' can be defined as $n' = \frac{1}{4}\sqrt{db_f}$, and the required baseplate thickness t_p can be calculated from

$$t_p = 2l\sqrt{\frac{f_p}{F_y}}$$

where $l = \max(m, n, n')$, in
$f_p = P/(BN) \leq F_p$, ksi
F_y = yield strength of baseplate, ksi
P = column axial load, kips

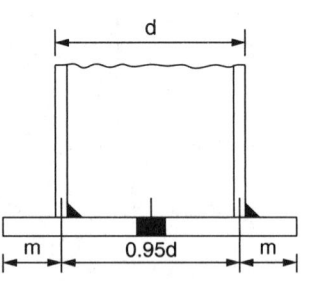

Figure 7.1 Column welded to a baseplate.

Plates: Bending Moments for Various Support and Loading Conditions

TABLE 7.3 Rectangular Plates: Bending Moments (Uniformly Distributed Load)

Plate Supports	b/a	α_a	α_b	β_a	β_b
4.	1.0	0.0334	0.0273	−0.0892	
	1.1	0.0349	0.0231	−0.0892	
	1.2	0.0357	0.0196	−0.0872	
	1.3	0.0359	0.0165	−0.0843	
	1.4	0.0357	0.0140	−0.0808	
	1.5	0.0350	0.0119	−0.0772	
	1.6	0.0341	0.101	−0.0735	
	1.7	0.0333	0.086	−0.0701	
	1.8	0.0326	0.0075	−0.0668	
	1.9	0.0316	0.0064	−0.0638	
	2.0	0.0303	0.0056	−0.0610	
5.	1.0	0.0273	0.0334		−0.0893
	1.1	0.0313	0.0313		−0.0867
	1.2	0.0348	0.0292		−0.0820
	1.3	0.0378	0.0269		−0.0760
	1.4	0.0401	0.0248		−0.0688
	1.5	0.0420	0.0228		−0.0620
	1.6	0.0433	0.0208		−0.0553
	1.7	0.0441	0.0190		−0.0489
	1.8	0.0444	0.0172		−0.0432
	1.9	0.0445	0.0157		−0.0332
	2.0	0.0443	0.0142		−0.0338
6.	1.0	0.0226	0.0198	−0.0556	−0.0417
	1.1	0.0234	0.0169	−0.0565	−0.0350
	1.2	0.0236	0.0142	−0.0560	−0.0292
	1.3	0.0235	0.0120	−0.0545	−0.0242
	1.4	0.0230	0.0102	−0.0526	−0.0202
	1.5	0.0225	0.0086	−0.0506	−0.0169
	1.6	0.0218	0.0073	−0.0484	−0.0142
	1.7	0.0210	0.0062	−0.0462	−0.0120
	1.8	0.0203	0.0054	−0.0442	−0.0102
	1.9	0.0192	0.0043	−0.0413	−0.0082
	2.0	0.0189	0.0040	−0.0404	−0.0076

NOTES

Local Buckling

Buckling may sometimes occur in the form of wrinkles in thin elements such as webs, flanges, cover plates, and other parts that make up a section. This phenomenon is called **local buckling**.

The critical buckling stress in rectangular plates with various of edge support and edge loading in the plane of the plates is given by

$$f_{cr} = k \frac{\pi^2 E}{12(1-\mu^2)(b/t)^2} \tag{7.2}$$

where k = constant that depends on the nature of loading, length-to-width ratio of plate, and edge conditions
E = modulus of elasticity
μ = Poisson's ratio
b = length of loaded edge of plate; or when the plate is subjected to shearing forces, the smaller lateral dimension
t = plate thickness

Table 7.1N lists values of k for various types of loads and edge support conditions. (From formulas, tables and curves in F Bleich, *Buckling Strength of Metal Structures*; S.P. Timoshenko and J. M. Gere, *Theory of Elastic Stability*; and G. Gernard, *Introduction to Structural Stability Theory*, McGraw-Hill, Inc., New York.)

TABLE 7.1N Values of k for Buckling Stress in Thin Plates

$\frac{a}{b}$	Case 1 (All edges clamped)	Case 2 (All edges simply supported)	Case 3 (Clamped edges)	Case 4 (Clamped edges)
0.4	28.3	8.4	9.4	
0.6	15.2	5.1	13.4	7.1
0.8	11.3	4.2	8.7	7.3
1.0	10.1	4.0	6.7	7.7
1.2	9.4	4.1	5.8	7.1
1.4	8.7	4.5	5.5	7.0
1.6	8.2	4.2	5.3	7.3
1.8	8.1	4.0	5.2	7.2
2.0	7.9	4.0	4.9	7.0
2.5	7.6	4.1	4.5	7.1
3.0	7.4	4.0	4.4	7.1
3.5	7.3	4.1	4.3	7.0
4.0	7.2	4.0	4.2	7.0
∞	7.0	4.0	4.0	∞

Plates: Bending Moments for Various Support and Loading Conditions

TABLE 7.4 Rectangular Plates: Bending Moments (Uniformly Distributed Load)

Plate Supports	b/a	α_a	α_b	β_a	β_b
7.	1.0	0.0180	0.0267		−0.0694
	1.1	0.0218	0.0262		−0.0708
	1.2	0.0254	0.0254		−0.0707
	1.3	0.0287	0.0242		−0.0689
	1.4	0.0316	0.0229		−0.0660
	1.5	0.0341	0.0214		−0.0621
	1.6	0.0362	0.0200		−0.0577
	1.7	0.0376	0.0186		−0.0531
	1.8	0.0388	0.0172		−0.0484
	1.9	0.0396	0.0158		−0.0439
	2.0	0.0400	0.0146		−0.0397
8.	1.0	0.0198	0.0226	−0.0417	−0.0556
	1.1	0.0226	0.0212	−0.0481	−0.0530
	1.2	0.0249	0.0198	−0.0530	−0.0491
	1.3	0.0266	0.0181	−0.0565	−0.0447
	1.4	0.0279	0.0162	−0.0588	−0.0400
	1.5	0.0285	0.0146	−0.0597	−0.0354
	1.6	0.0289	0.0130	−0.0599	−0.0312
	1.7	0.0290	0.0116	−0.0594	−0.0274
	1.8	0.0288	0.0103	−0.0583	−0.0240
	1.9	0.0284	0.0092	−0.0570	−0.0212
	2.0	0.0280	0.0081	−0.0555	−0.0187
9.	1.0	0.0179	0.0179	−0.0417	−0.0417
	1.1	0.0194	0.0161	−0.0450	−0.0372
	1.2	0.0204	0.0142	−0.0468	−0.0325
	1.3	0.0208	0.0123	−0.0475	−0.0281
	1.4	0.0210	0.0107	−0.0473	−0.0242
	1.5	0.0208	0.0093	−0.0464	−0.0206
	1.6	0.0205	0.0080	−0.0452	−0.0177
	1.7	0.0200	0.0069	−0.0438	−0.0152
	1.8	0.0195	0.0060	−0.0423	−0.0131
	1.9	0.0190	0.0052	−0.0408	−0.0113
	2.0	0.0183	0.0046	−0.0392	−0.0098

--- NOTES ---

Bearing Plates

To resist a beam reaction, the minimum bearing length N in the direction of the beam span for a bearing plate is determined by equations for prevention of local web yielding and web crippling. A larger N is generally desirable but is limited by the available wall thickness.

When the plate covers the full area of a concrete support, the area, in², required by the bearing plate is

$$A_1 = \frac{R}{0.35 f'_c} \qquad (7.3)$$

where R = beam reaction, kips
f'_c = specified compressive strength of the concrete, ksi

When the plate covers less than the full area of the concrete support, then, as determined from Table 7.2N,

$$A_1 = \left(\frac{R}{0.35 f'_c \sqrt{A_2}}\right)^2 \qquad (7.4)$$

where A_2 = full cross-sectional area of concrete support, in²
With N established, usually rounded to full inches, the minimum width of plate B, in, may be calculated by dividing A_1 by N and then rounding off to full inches so that $BN \geq A_1$. Actual bearing pressure f_p, ksi, under the plate then is

$$f_p = \frac{R}{BN} \qquad (7.5)$$

The plate thickness usually is determined with the assumption of cantilever bending of the plate.

$$t = \left(\frac{1}{2}B - k\right)\sqrt{\frac{3f_p}{F_b}} \qquad (7.6)$$

where t = minimum plate thickness, in
k = distance, in, from beam bottom to top of web fillet
F_b = allowable bending stress of plate, ksi

TABLE 7.2N Allowable Bearing Stress F_p on Concrete and Masonry, ksi

Full area of concrete support	$0.35 f'_c$
Less than full area of concrete support	$0.35 f'_c \sqrt{A_1/A_2} \leq 0.70 f'_c$
Sandstone and limestone	0.40
Brick in cement mortar	0.25

Plates: Bending Moments for Various Support and Loading Conditions

TABLE 7.5 Rectangular Plates Bending Moments (Uniformly Distributed Load)

Plate Supports	b/a	α_a	α_b	β_a	β_b
10.	1.0	0.0099	0.0457	−0.0510	−0.0853
	1.1	0.0102	0.0492	−0.0574	−0.0930
	1.2	0.0102	0.0519	−0.0636	−0.1000
	1.3	0.0100	0.0540	−0.0700	−0.1062
	1.4	0.0097	0.00552	−0.0761	−0.1115
	1.5	0.0095	0.0556	−0.0821	−0.1155
11.	1.0	0.0457	0.0099	−0.0853	−0.0510
	1.1	0.0421	0.0094	−0.0777	−0.0448
	1.2	0.0389	0.0087	−0.0712	−0.0397
	1.3	0.0362	0.0079	−0.0658	−0.0354
	1.4	0.0362	0.0070	−0.0609	−0.0314
	1.5	0.0311	0.0059	−0.0562	−0.0279

Bending Moments (Concentrated Load at Center)

$$M_{0(a)} = \alpha_a \cdot P, \quad M_{0(b)} = \alpha_b \cdot P, \quad M_{s(a)} = \beta_a \cdot P, \quad M_{s(b)} = \beta_b \cdot P$$

Plate Supports	b/a	α_a	α_b	β_a	β_b
1.	1.0	0.146	0.146		
	1.2	0.179	0.141		
	1.4	0.214	0.138		
	1.6	0.244	0.135		
	1.8	0.270	0.132		
	2.0	0.290	0.130		
2.	1.0	0.108	0.108	−0.094	−0.094
	1.2	0.128	0.100	−0.126	−0.074
	1.4	0.143	0.092	−0.149	−0.055
	1.6	0.156	0.086	−0.162	−0.040
	1.8	0.162	0.080	−0.171	−0.030
	2.0	0.168	0.076	−0.176	−0.022

--- NOTES ---

Example. Computation of rectangular plate, $b \leq 2a$

Given. Elastic plate 1 in Table 7.6, $a = 1.8$ m, $b = 2.25$ m, $t = 0.1$ m, $a/b = 0.8$

Modulus of elasticity $E = 4030$ kips/in$^2 = \dfrac{4030 \times 4.48222}{2.54^2} = 2800$ kN/cm^2

Poisson's ratio $\mu = \mu_T = 1/6$,

Elastic stiffness $D = \dfrac{Et^3}{12(1-\mu^2)} = \dfrac{2800 \times 10^3}{12\left[1-(1/6)^2\right]} = 24,0000$

Uniformly distributed load $w = 0.2$ kN/m$^2 = 0.002$ kN/cm^2

Required. Compute bending moments $M_{0(a)}$ and $M_{0(b)}$, deflection Δ_0.

Solution. $M_{0(a)} = \alpha_a w b^2 = 0.0323 \times 0.2 \times 2.25^2 = 0.0327$ kN·m/m $= 32.7$ N·m/m

$M_{0(b)} = \alpha_b w b^2 = 0.1078 \times 0.2 \times 2.25^2 = 0.1091$ kN·m/m $= 109.1$ N·m/m

$\Delta_0 = \eta_0 w \dfrac{b^4}{D} = 0.018 \times 0.002 \times \dfrac{225^4}{24,000} = 0.38$ cm $= 3.8$ mm

Plates: Bending Moments for Various Support and Loading Conditions

TABLE 7.6 Rectangular Plates: Bending Moments and Deflections (Uniformly Distributed Load)

$$M_{0(a)} = \alpha_a \cdot w \cdot b^2, \quad M_{0(b)} = \alpha_b \cdot w \cdot b^2, \quad M_{1(a)} = \alpha_{1(a)} \cdot w \cdot b^2, \quad M_{2(b)} = \alpha_{2(b)} \cdot w \cdot b^2$$

$\alpha_a, \alpha_b, \alpha_{1(a)},$ and $\alpha_{2(b)}$ = coefficients for Poisson's ratio $\mu_T = 1/6$

$$\Delta_0 = \eta_0 \cdot w \cdot \frac{b^4}{D}, \quad \Delta_i = \eta_1 \cdot w \cdot \frac{b^4}{D}, \quad \Delta_2 = \eta_2 \cdot w \cdot \frac{b^4}{D}, \quad D = \frac{E \cdot t^3}{12(1-\mu^2)}$$

where Δ_i = deflection at point i
t = plate thickness
E = modulus of elasticity
μ = poisson's ratio
D = elastic stiffness

Plate Supports	a/b	$\alpha_{0(a)}$	$\alpha_{0(b)}$	$\alpha_{1(a)}$	$\alpha_{2(a)}$	η_0	η_1	η_2
1.	1.0	0.0947	0.0947	0.1606	0.1606	0.0263	0.0172	0.0172
	0.9	0.0689	0.1016	0.1367	0.1541	0.0218	0.0119	0.0164
	0.8	0.0479	0.1078	0.1148	0.1486	0.0180	0.0079	0.0157
	0.7	0.0289	0.1132	0.0955	0.1435	0.0158	0.0050	0.0151
	0.6	0.0131	0.1178	0.0769	0.1386	0.0148	0.0030	0.0146
	0.5	0.0005	0.1214	0.0592	0.1339	0.0140	0.0016	0.0141
2.	1.0	0.0977	0.1070	0.1578	0.2326	0.0606	0.0168	0.1011
	0.9	0.1007	0.0889	0.1552	0.2073	0.0418	0.0165	0.0625
	0.8	0.1038	0.0729	0.1526	0.1844	0.0307	0.0162	0.0406
	0.7	0.1069	0.0589	0.1498	0.1639	0.0247	0.0159	0.0275
	0.6	0.1097	0.0468	0.1470	0.1462	0.0209	0.155	0.0194
	0.5	0.1121	0.0364	0.1444	0.1314	0.185	0.0152	0.0142
3.	1.0	0.0581	0.0581	0.1198	0.1198	0.0122	0.0126	0.0126
	0.9	0.0500	0.0540	0.1031	0.1092	0.0100	0.0089	0.0117
	0.8	0.0421	0.0490	0.0866	0.0986	0.0080	0.0059	0.0106
	0.7	0.0343	0.0432	0.0706	0.0870	0.0063	0.0037	0.0093
	0.6	0.0270	0.0367	0.0547	0.0739	0.0048	0.0022	0.0078
	0.5	0.0202	0.0294	0.0388	0.0578	0.0036	0.0011	0.0063

Flange Plate Thickness

The flange of a welded plate girder is actually a series of plates which are married to one another, end to end, using full penetration butt welds. A full penetration butt weld implies a weld where the flange plates come together at a *butt joint* (i.e., where two pieces of steel lie approximately in the same horizontal plane). The weld, also known as a *complete penetration groove weld*, extends completely through the flanges to be joined, is designed to transmit the total load and has the same strength as that of the flange plates. This type of weld is much more difficult to fabricate than the fillet welds discussed earlier in the section on cover plate design. Compared to a fillet weld, though, the groove weld provides for a stronger connection.

The ratio of the compression flange plate width to its thickness should be specified so that it does not exceed (AASHTO 10.34.2.1.3)

$$\frac{b}{t} = \frac{3250}{\sqrt{f_b}} \leq 24 \tag{7.7}$$

where b = width of flange plate
t = thickness of flange plate
f_b = computed compressive bending stress

When a hybrid, grider is being used, computed compressive bending stress f_b is divided by 3 reduction factor R. This reduction factor, which is utilized throughout the design of a hybrid girder, is defined as (AASHTO 10.40.2.1)

$$R = 1 - \frac{\beta\psi(1-\alpha)^2(3-\psi+\psi\alpha)}{6+\beta\psi(3-\psi)} \tag{7.8}$$

where $\alpha = \dfrac{\text{minimum yield strength of web}}{\text{minimum yield strength of tension flange}}$

$\beta = \dfrac{\text{area of web}}{\text{area of tension flange}}$

$\psi = \dfrac{\text{distance outer edge of tension flange to N.A.}}{\text{depth of steel section}}$

Plates: Bending Moments for Various Support and Loading Conditions

TABLE 7.7 Rectangular Plates: Bending Moments (Uniformly Varying Load)

$$M_{0(a)} = \alpha_a \cdot w \cdot \left(\frac{a \cdot b}{2}\right), \quad M_{0(b)} = \alpha_b \cdot w \cdot \left(\frac{a \cdot b}{2}\right), \quad M_{s(a)} = \beta_a \cdot w \cdot \left(\frac{a \cdot b}{2}\right), \quad M_{s(b)} = \beta_b \cdot w \cdot \left(\frac{a \cdot b}{2}\right)$$

Plate Supports	b/a	α_a	α_b	β_a	β_b
1.	1.0	0.0216	0.0194	−0.0502	−0.0588
	1.1	0.0229	0.0178	−0.0515	−0.0554
	1.2	0.0236	0.0161	−0.0521	−0.0517
	1.3	0.0239	0.0145	−0.0522	−0.0477
	1.4	0.0241	0.0131	−0.0519	−0.0432
	1.5	0.0241	0.0117	−0.0514	−0.0387
2.	1.0	0.0194	0.0216	−0.0588	−0.0502
	1.1	0.0211	0.0198	−0.0614	−0.0480
	1.2	0.0228	0.0178	−0.0633	−0.0435
	1.3	0.0243	0.0153	−0.0644	−0.0418
	1.4	0.0257	0.0132	−0.0650	−0.0396
	1.5	0.0271	0.0120	−0.0652	−0.0357
3.	1.0	0.0246	0.0172	−0.0538	−0.0598
	1.1	0.0248	0.0163	−0.0538	−0.0553
	1.2	0.0250	0.0153	−0.0535	−0.0510
	1.3	0.0250	0.0142	−0.0529	−0.0469
	1.4	0.0247	0.0128	−0.0522	−0.0429
	1.5	0.0245	0.0114	−0.0514	−0.0390
4.	1.0	0.0172	0.0246	−0.0598	−0.0538
	1.1	0.0178	0.0244	−0.0640	−0.0535
	1.2	0.0180	0.0242	−0.0677	−0.0533
	1.3	0.0182	0.0244	−0.0709	−0.0533
	1.4	0.0180	0.0249	−0.0739	−0.0536
	1.5	0.0177	0.0262	−0.0765	−0.555

--- NOTES ---

Stresses in Plates

Circular Plate, Uniform Load, Simply Supported Edges
The maximum stress occurs in the center and is

$$f = (3wr^2 \times 3.30)/8t^2 \tag{7.9}$$

Circular Plate, Uniform Load, Fixed Edges
The maximum stress occurs at the edge and is

$$f = 3wr^2/4t^2 \tag{7.10}$$

Rectangular Plate, Uniform Load, Simply Supported Edges
The maximum stress occurs at the center and is

$$f = 3wb^2/4t^2(1 + 1.61b^2/a^2) \tag{7.11}$$

Rectangular Plate, Uniform Load, Fixed Edges
The maximum stress occurs at the center of the long edge and is

$$f = wb^2/2t^2(1 + 0.623b^6/a^6) \tag{7.12}$$

Equations (7.9) to (7.12) are based on a value of Poisson's ratio of 0.30. The nomenclature for the equations is

f = maximum stress, psi
w = uniform load, psi
r = radius of circular plate, in
b = short dimension of rectangular plate, in
a = long dimension of rectangular plate, in
t = thickness of plate, in

Plates: Bending Moments for Various Support and Loading Conditions

TABLE 7.8 Rectangular Plates: Bending Moments (Uniformly Varying Load)

$$M_{0(a)} = \alpha_a \cdot w \cdot \left(\frac{a \cdot b}{2}\right), \quad M_{0(b)} = \alpha_b \cdot w \cdot \left(\frac{a \cdot b}{2}\right), \quad M_{s(a)} = \beta_a \cdot w \cdot \left(\frac{a \cdot b}{2}\right), \quad M_{s(b)} = \beta_b \cdot w \cdot \left(\frac{a \cdot b}{2}\right)$$

Plate Supports	b/a	α_a	α_b	β_b	β_c
5.	1.0	0.0718	0.0042	−0.1412	−0.0422
	1.1	0.0672	0.0037	−0.1308	−0.0350
	1.2	0.0634	0.0031	−0.1222	−0.0290
	1.3	0.0598	0.0025	−0.1143	−0.0240
	1.4	0.0565	0.0019	−0.1069	−0.0200
	1.5	0.0530	0.0012	−0.1003	−0.0168
6.	1.0	0.0042	0.0718	−0.0422	−0.1412
	1.1	0.0047	0.0758	−0.0509	−0.1510
	1.2	0.0053	0.0790	−0.0600	−0.1600
	1.3	0.0057	0.0810	−0.0692	−0.1675
	1.4	0.0060	0.0826	−0.0785	−0.1740
	1.5	0.0063	0.0828	−0.0876	−0.1790

$$M_{0(a)} = \alpha_a \cdot w \cdot \left(\frac{a \cdot b}{2}\right), \quad M_{0(b)} = \alpha_b \cdot w \cdot \left(\frac{a \cdot b}{2}\right)$$

$$M_{s(1)} = \beta_1 \cdot w \cdot \left(\frac{a \cdot b}{2}\right), \quad M_{s(2)} = \beta_2 \cdot w \cdot \left(\frac{a \cdot b}{2}\right), \quad M_{s(3)} = \beta_3 \cdot w \cdot \left(\frac{a \cdot b}{2}\right)$$

Plate Supports	b/a	α_a	α_b	β_1	β_2	β_3
7.	1.0	0.0184	0.0206	−0.0448	−0.0562	−0.0332
	1.1	0.0205	0.0190	−0.0477	−0.0538	−0.0302
	1.2	0.0221	0.0173	−0.0495	−0.0506	−0.0271
	1.3	0.0229	0.0156	−0.0504	−0.0470	−0.0237
	1.4	0.0235	0.0137	−0.0508	−0.0431	−0.0204
	1.5	0.0241	0.0120	−0.0510	−0.0387	−0.0168
8.	1.0	0.0206	0.0184	−0.0562	−0.0332	−0.0446
	1.1	0.0218	0.0160	−0.0576	−0.0353	−0.0411
	1.2	0.0227	0.0137	−0.0580	−0.0357	−0.0372
	1.3	0.0231	0.0112	−0.0577	−0.0376	−0.0336
	1.4	0.0233	0.0090	−0.0569	−0.0380	−0.0302
	1.5	0.0233	0.0072	−0.0556	−0.0382	−0.0276

NOTES

TABLE 7.3N Formulas for Flat Plates*

Notation: W = total applied load (lb); w = unit applied load (lb per sq. in); t = thickness of plate (in); s = unit stress at surface of plate (lb per sq. in); y = vertical deflection of plate from original position (in); θ = slope of plate measured from horizontal (rad); E = modulus of elasticity; m = reciprocal of v, Poisson's ratio. q denotes any given point on the surface of plate; r denotes the distance of q from the center of a circular plate. Other dimensions and corresponding symbols are indicated on figures. Positive sign for s indicates tension at upper surface and equal compression at lower surface; negative sign indicates reverse conditions. Positive sign for y indicates upward deflection, negative sign downward deflection. Subscripts r, t, a, and b used with s denote, respectively, radial direction, tangential direction, direction of dimension a, and direction of dimension b. All dimensions are in inches. All logarithms are to the base e, ($\log_{10} x = 2.3026 \log_{10} x$).

Manner of Loading and Case No.	Formulas for Stress and Deflection
Edges supported 1. Uniform load over entire surface $W = \omega\pi a^2$	Circular and solid (A) (At q) $s_r = -\dfrac{3W}{8\pi m l^2}\left[(3m+1)\left(1-\dfrac{r^2}{a^2}\right)\right]$ $s_t = -\dfrac{3W}{8\pi m l^2}\left[(3m+1)-(m+3)\dfrac{r^2}{a^2}\right]$ $y = -\dfrac{3W(m^2-1)}{8\pi E m^2 l^2}\left[\dfrac{(5m+1)a^2}{2(m+1)}+\dfrac{r^4}{2a^2}-\dfrac{(3m+1)r^2}{m+1}\right]$ (At center) max $s_r = s_t = -\dfrac{3W}{8\pi m l^2}(3m+1)$ max $y = -\dfrac{3W(m-1)(5m+1)a^2}{16\pi E m^2 l^3}$ (At edge) $\theta = \dfrac{3W(m-1)a}{2\pi E m l^3}$

*From: Roark, *Formulas for Stress and Strain*, 4th ed., McGraw-Hill.

Continued on page 146

Plates: Bending Moments for Various Support and Loading Conditions 145

TABLE 7.9 Circular Plates: Bending Moments, Shear and Deflection (Uniformly Distributed Load)

a = circular plate's radius r = circular section's radius t = thickness of plate	M_R = radial moment M_T = tangential moment V_R = radial shear R = support reaction Δ = deflection at center of plate μ = Poisson's ratio E = modulus of elasticity

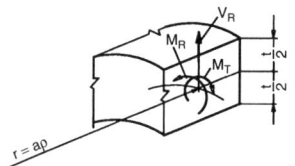

Moment, Shear and Deflection Diagrams	Formulas
1. 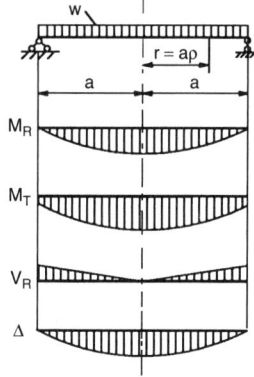	$\rho = \dfrac{r}{a}, \quad P = w\pi a^2, \quad R = \dfrac{P}{2\pi a}, \quad V_R = \dfrac{P}{2\pi a}\rho$ $M_R = \dfrac{P}{16\pi}(3+\mu)(1-\rho^2)$ $M_T = \dfrac{P}{16\pi}\left[3+\mu-(1+3\mu)\rho^2\right]$ $\Delta = \dfrac{Pa^2}{64\pi D}(1-\rho^2)\left(\dfrac{5+\mu}{1+\mu}-\rho^2\right), \quad D = \dfrac{Et^3}{12(1-\mu^2)}$
2. 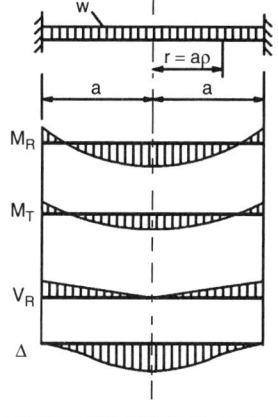	$\rho = \dfrac{r}{a}, \quad P = W\pi a^2, \quad R = \dfrac{P}{2\pi a}, \quad V_R = -\dfrac{P}{2\pi a}\rho$ $M_R = \dfrac{P}{16\pi}[1+\mu-(3+\mu)\rho^2]$ $M_1 = \dfrac{P}{16\pi}[1+\mu-(1+3\mu)\rho^2]$ $\Delta = \dfrac{Pa^2}{64\pi D}(1-\rho^2), \quad D = \dfrac{Et^3}{12(1-\mu^2)}$

NOTES

TABLE 7.3N Formulas for Flat Plates* *(Continued)*

Manner of Loading and Case No.	Formulas for Stress and Deflection
Edges supported 2. Uniform load over concentric circular area of radius r_o $W = \omega\pi r_o^2$	(At $q, r < r_o$) $s_r = -\dfrac{3W}{2\pi ml^2}\left[m + (m+1)\log\dfrac{a}{r_o} - (m-1)\dfrac{r_o^2}{4a^2} - (3m+1)\dfrac{r^2}{4r_o^2}\right]$ $s_t = -\dfrac{3W}{2\pi ml^2}\left[m + (m+1)\log\dfrac{a}{r_o} - (m-1)\dfrac{r_o^2}{4a^2} - (m+3)\dfrac{r^2}{4r_o^2}\right]$ $y = -\dfrac{3W(m^2-1)}{16\pi Em^2 l^3}\left[\begin{array}{l}4a^2 - 5r_o^2 + \dfrac{r^4}{r_o^2} - (8r^2 + 4r_o^2)\log\dfrac{a}{r_o} - \\ \dfrac{2(m-1)r_o^2(a^2-r^2)}{(m+1)a^2} + \dfrac{8m(a^2-r^2)}{m+1}\end{array}\right]$ (At $q, r > r_o$) $s_r = -\dfrac{3W}{2\pi ml^2}\left[(m+1)\log\dfrac{a}{r} - (m-1)\dfrac{r_o^2}{4a^2} + (m-1)\dfrac{r_o^2}{4r^2}\right]$ $s_t = -\dfrac{3W}{2\pi ml^2}\left[(m-1) + (m+1)\log\dfrac{a}{r} - (m-1)\dfrac{r_o^2}{4a^2} - (m-1)\dfrac{r_o^2}{4r^2}\right]$ $y = -\dfrac{3W(m^2-1)}{16\pi Em^2 l^3}\left[\begin{array}{l}\dfrac{(12m+4)(a^2-r^2)}{m+1} - \\ \dfrac{2(m-1)r_o^2(a^2-r^2)}{(m+1)a^2} - (8r^2 + 4r_o^2)\log\dfrac{a}{r}\end{array}\right]$ (At center) Max $s_r = s_t = -\dfrac{3W}{2\pi ml^2}\left[m + (m+1)\log\dfrac{a}{r_o} - (m-1)\dfrac{r_o^2}{4a^2}\right]$ Max $y = -\dfrac{3W(m^2-1)}{16\pi Em^2 l^3}\left[\dfrac{(12m+4)a^2}{m+1} - 4r_o^2\log\dfrac{a}{r_o} - \dfrac{(7m+3)r_o^2}{m+1}\right]$ For r_o very small (concentrated load) Max $y = -\dfrac{3W(m-1)(3m+1)a^2}{4\pi Em^2 l^3}$ (At edge) $\theta = -\dfrac{3W(m-1)a}{\pi Em l^3}$

*From: Roark, *Formulas for Stress and Strain*, 4th ed., McGraw-Hill.

Plates: Bending Moments for Various Support and Loading Conditions

--- NOTES ---

TABLE 7.3N Formulas for Flat Plates* *(Continued)*

Manner of Loading and Case No.	Formulas for Stress and Deflection
Outer edge supported 3. Uniform load over entire actual surface $W = \omega\pi(a^2 - b^2)$	Circular, with concentric circular hole (circular flange) (B) (At inner edge) Max $s = s_t$ $$= -\frac{3w}{4ml^2(a^2-b^2)}\left[a^4(3m+1) + b^4(m-1) - 4ma^2b^2 - 4(m+1)a^2b^2\log\frac{a}{b}\right]$$ When b is very small, $\text{Max } s = s_t = -\dfrac{3wa^2(3m+1)}{4ml^2}$ $$\text{Max } y = -\frac{3w(m^2-1)}{2m^2El^3}\left[\frac{a^4(5m+1)}{8(m+1)} + \frac{b^4(7m+3)}{8(m+1)} - \frac{a^2b^2(3m+1)}{2(m+1)} + \frac{a^2b^2(3m+1)}{2(m-1)}\log\frac{a}{b} - \frac{2a^2b^4(m+1)}{(a^2-b^2)(m-1)}\left(\log\frac{a}{b}\right)^2\right]$$
Outer edge supported 4. Uniform load along inner edge $\downarrow W \downarrow$	(At inner edge) $$\text{Max } s = s_t = -\frac{3W}{2\pi^2 ml^2}\left[\frac{2a^2(m+1)}{a^2-b^2}\log\frac{a}{b} + (m-1)\right]$$ $$\text{Max } y = -\frac{3W(m^2-1)}{4\pi Em^2l^3}\left[\frac{(a^2-b^2)(3m+1)}{(m+1)} + \frac{4a^2b^2(m+1)}{(m-1)(a^2-b^2)}\left(\log\frac{a}{b}\right)^2\right]$$
Supported along concentric circle near outer edge 5. Uniform load along concentric circle near inner edge.	(At inner edge) $$\text{Max } s = s_t = -\frac{3W}{2\pi ml^2}\left[\frac{2a^2(m+1)}{a^2-b^2}\log\frac{c}{d} + (m-1)\frac{c^2-d^2}{a^2-b^2}\right]$$

PART III
Soils and Foundations

| **CHAPTER 8**
Soils | **CHAPTER 9**
Foundations |

CHAPTER 8
Soils

Soils and Foundations

NOTES

For purposes of structural design, engineering properties of soils are determined through laboratory experiments and field research, conducted for specific conditions. If these methods are unavailable, use of data provided in the norms may be acceptable.

The modulus of deformation and Poisson's ratio of soil can be determined using the following formulas:

$$E_s = \frac{3c_1 c_2}{2c_1 + c_2}, \qquad \mu = \frac{c_1 - c_2}{2c_1 + c_2}$$

$$c_1 = \frac{(1+2k_0)(1+e)}{D_r}, \qquad c_2 = \frac{(1-k_0)(1+e)}{D_r}$$

where k_0 = coefficient of lateral earth pressure (Table 10.1)
e = void ratio (Table 8.2)
D_r = relative density (Table 8.2)

Soil properties found in Tables 8.2 to 8.7 are provided only as guidelines.

TABLE 8.1 Soils: Engineering Properties

Soil Type	Soil Particles	
	Size	Weight in Dry Soil
Cohesive soils Igneous and sedimentary stone compact soils; compact, sticky, and plastic clay soils	Less than 0.005 mm	
Cohesionless soils Crashed stone	Coarser than 10 mm	> 50%
Gravel sand	Coarser than 2 mm	> 50%
Coarse-grained sand	Coarser than 0.5 mm	> 50%
Medium-grained sand	Coarser than 0.25 mm	> 50%
Fine-grained sand	Coarser than 0.1 mm	> 75%
Dustlike sand	Coarser than 0.1 mm	< 75%

Components of Soil

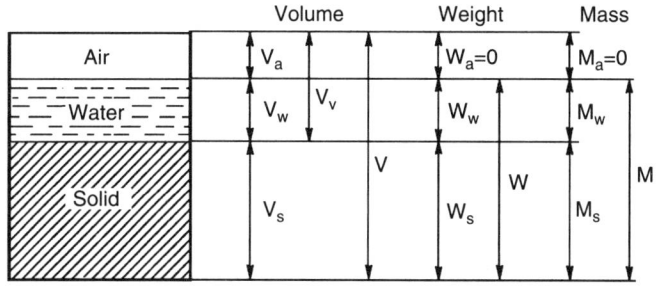

V, V_a, V_w, V_s, and V_v = total volume and volume of air, water, solid matter, and voids, respectively
W, W_w, and W_s = total weight and weight of water and solid matter, respectively
M, M_w, and M_s = total mass and mass of water and solid matter, respectively

NOTES

Relationship of Weights and Volumes in Soil

The unit weight of soil varies, depending on the amount of water contained in the soil. Three unit weights are in general use: the saturated unit weight γ_{sat}, the dry unit weight γ_{dry}, and the buoyant unit weight γ_b.

$$\gamma_{sat} = (G+e)\gamma_0/(1+e) = (1+w)G\gamma_0/(1+e) \qquad S = 100\% \qquad (8.1)$$
$$\gamma_{dry} = G\gamma_0/(1+e) \qquad\qquad\qquad\qquad\qquad S = 0\% \qquad (8.2)$$
$$\gamma_b = (G-1)\gamma_0/(1+e) \qquad\qquad\qquad\qquad S = 100\% \qquad (8.3)$$

Unit weights are generally expressed in pounds per cubic foot or grams per cubic centimeter. Representative values of unit weights for a soil with a specific gravity of 2.73 and a void ratio of 0.80 are

$$\gamma_{sat} = 122 \text{ lb/ft}^3 = 1.96 \text{ g/cm}^3$$
$$\gamma_{dry} = 95 \text{ lb/ft}^3 = 1.52 \text{ g/cm}^3$$
$$\gamma_b = 60 \text{ lb/ft}^3 = 0.96 \text{ g/cm}^3$$

The symbols used in Eqs. (8.1) and (8.2) and in Fig. 8.1 are

G = specific gravity of soil solids (specific gravity of quartz is 2.67; for majority of soils, specific gravity ranges between 2.65 and 2.85; organic soils would have lower specific gravities)

γ_0 = unit weight of water (62.4 lb/ft^3 or 1.0 g/cm^3)

e = voids ratio, volume of voids in mass of soil divided by volume of solids in same mass (also equal to $n/(1-n)$, where n is porosity—volume of voids in mass of soil divided by total volume of same mass)

S = degree of saturation, volume of water in mass of soil divided by volume of voids in same mass

w = water content, weight of water in mass of soil divided by weight of solids in same mass (also equal to Se/G)

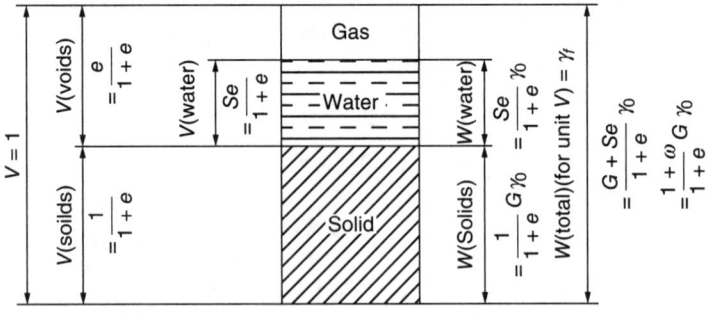

FIGURE 8.1 Relationship of weights and volumes in soil.

Soils

TABLE 8.2 Soils: Weight/Mass and Volume Relationships

1. Porosity: $n = \dfrac{V_v}{V} \cdot 100\%$, $V = V_s + V_v$	9. **Specific gravity of solids:** $$G_s = \dfrac{W_s/V_s}{\gamma_w} = \dfrac{W_s}{V_s \cdot \gamma_w} \text{ or }$$ $$G_s = \dfrac{M_s/V_s}{\rho_w} = \dfrac{M_s}{V_s \cdot \rho_w}$$ where γ_w and ρ_w = unit weight and unit mass of water γ_w = 62.4 lb/ft³ or 9.81 kN/m³, ρ_w = 1000 kg/m³ (at normal temperatures)
2. Void ratio: $e = \dfrac{V_v}{V_s} = \dfrac{n}{1-n}$, $V_v = V_a + V_w$	
3. Degree of saturation: $S = \dfrac{V_w}{V_v} \cdot 100\%$	
4. Water content: $w = \dfrac{W_w}{W_s} \cdot 100\% = \dfrac{M_w}{M_s} \cdot 100\%$	
5. Unit weight: $\gamma = \dfrac{W_s + W_w}{V}$	
6. Dry unit weight: $\gamma_d = \dfrac{W_s}{V} = \dfrac{\gamma}{1+w}$	10. **Relative density:** $$D_r = \dfrac{e_{max} - e_0}{e_{max} - e_{min}} \cdot 100\%$$ or $D_r = \dfrac{\gamma_{max}(\gamma - \gamma_{min})}{\gamma(\gamma_{max} - \gamma_{min})} \cdot 100\%$ where e_{max}, e_{min}, and e_0 = maximum, minimum, and in-place void ratio of the soil, respectively. γ_{max}, γ_{min}, and γ_0 = maximum, minimum, and in-place dry unit weight, respectively.
7. Unit mass: $\rho = \dfrac{M}{V}$	
8. Dry unit mass: $\rho_d = \dfrac{M_s}{V}$	

Flow of Water in Soil		
Darcy's law. Velocity of flow: $v = k_p \cdot i$, where k_p = coefficient of permeability $I = \dfrac{\Delta H}{\Delta L}$ = hydraulic gradient (slope) Actual velocity: $v_{actual} = \dfrac{v}{n} = \dfrac{k_p \cdot i}{n}$ or $v_{actual} = \dfrac{k_p \cdot i(1+e)}{e}$ where n and e = soil's porosity and void ratio, respectively Flow rate (volume per unit time): $q = k_p \cdot i \cdot A$ where A = area of the given cross section of soil.	**Coefficient of Permeability (k_p)**	
	Soil Type	k_p, cm/s
	Crashed stone, gravel sand	1×10^{-1}
	Coarse-grained sand	1×10^{-2} to 1×10^{-1}
	Medium-grained sand	1×10^{-3} to 1×10^{-2}
	Fine-grained sand	1×10^{-4} to 1×10^{-3}
	Sandy loam	1×10^{-5} to 1×10^{-3}
	Sandy clay	1×10^{-7} to 1×10^{-5}
	Clay	$< 10^{-7}$

─── NOTES ───

Lateral Pressures in Soils, Forces on Retaining Walls

The Rankine theory of lateral earth pressures, used for estimating approximate values for lateral pressures on retaining walls, assumes that the pressure on the back of a vertical wall is the same as the pressure that would exist on a vertical plane in an infinite soil mass. Friction between the wall and the soil is neglected. The pressure on a wall consists of (1) the lateral pressure of the soil held by the wall, (2) the pressure of the water (if any) behind the wall, and (3) the lateral pressure from any surcharge on the soil behind the wall.

Symbols used in this section are as follows:

γ = unit weight of soil, lb/ft³ (kg/m³) (saturated unit weight, dry unit weight, or buoyant unit weight, depending on conditions)
P = total thrust of soil, lb/linear ft (kg/linear m) of wall
H = total height of wall, ft (m)
ϕ = angle of internal friction of soil, deg
i = angle of inclination of ground surface behind wall with horizontal; also angle of inclination of line of action of total thrust P and pressures on wall with horizontal
K_A = coefficient of active pressure
K_p = coefficient of passive pressure
c = cohesion, lb/ft² (kPa)

Vertical Pressures in Soils

The vertical stress in a soil caused by a vertical, concentrated surface load may be determined with a fair degree of accuracy by the use of elastic theory. Two equations are in common use, the Boussinesq and the Westergaard. The Boussinesq equation applies to an elastic, isotropic, homogeneous mass that extends infinitely in all directions from a level surface. The vertical stress at a point in the mass is

$$\sigma_z = \frac{3P}{2\pi z^2}\left[1+\left(\frac{r}{z}\right)^2\right]^{5/2}$$

The Westergaard equation applies to an elastic material laterally reinforced with horizontal sheets of negligible thickness and infinite rigidity, which prevent the mass from undergoing lateral strain. The vertical stress at a point in the mass, assuming a Poisson's ratio of zero, is

$$\sigma_z = \frac{P}{\pi z^2}\left[1+2\left(\frac{r}{z}\right)^2\right]^{3/2}$$

where σ_z = vertical stress at a point, lb/ft² (kPa)
P = total concentrated surface load, lb (N)
z = depth of point at which σ_z acts, measured vertically downward from surface, ft (m)
r = horizontal distance from projection of surface load P to point at which σ_z acts ft (m)

Continued on page 158

TABLE 8.3 Soils: Stress Distribution

Method Based on Elastic Theory

Concentrated Load

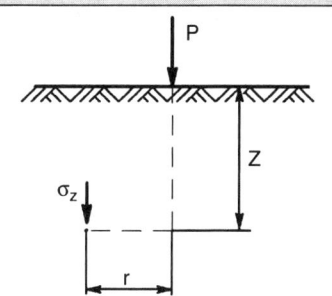

Boussinesq equation:

$$\sigma_z = \frac{3P}{2\pi z^2 [1+(r/z)^2]^{5/2}}$$

where σ_z = vertical stress at depth z
P = concentrated load

Uniformly Distributed Load

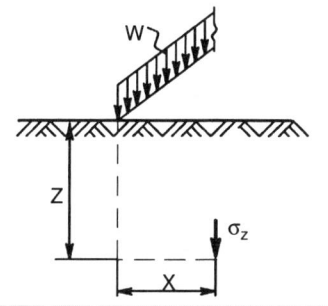

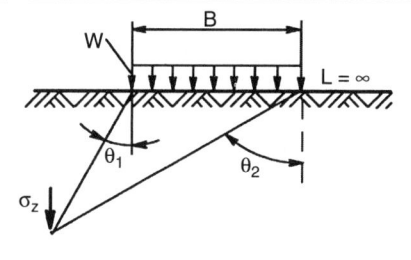

$$\sigma_z = \frac{2w}{\pi z[1+(x/z)^2]^2}$$

$$\sigma_z = \frac{w}{\pi}(\theta_2 - \theta_1 + \sin\theta_2\cos\theta_2 - \sin\theta_1\cos\theta_1)$$

Approximate Method

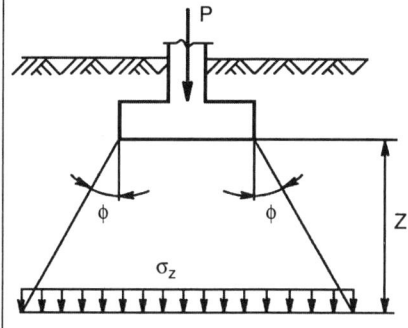

$$\sigma_z = \frac{P}{(B+2z\tan\phi)(L+2z\tan\phi)}$$

where σ_z = approximate vertical stress at depth z
P = total load
B = width of footing
L = length of footing, $B < L$
z = depth
ϕ = angle of internal friction

NOTES

For value of r/z between 0 and 1, the Westergaad equation gives stresses appreciably lower than those given by the Boussinesq equation. For values of r/z greater than 2.2, both equations give stresses less than $P/100z^2$.

Example for Table 8.4. Settlement of soil. Method based on elastic theory.

Units: $B(m)$, $L(m)$, $H(m)$, $h_i(m)$, $P_v(kN)$, $\gamma_i(kN/m^3)$, $\sigma_{a_i}(kPa)$, $E_{s_i}(kPa)$

P_v = weight of structures + weight of footing and surcharge + temporary load (live load)

z_i = distance from footing base to the middle of h_i layer

Lower border of active soil zone for vertical load P_v has been adopted as 20% of natural soil pressure: $0.2\sigma_\gamma$.

Given. $B = 3$ m, $L = 5.4$ m, $H_1 = 5$ m, $h_0 = 2$ m, $h_1 = h_2 = h_3 = 1.0$ m $< 0.4B$

$H_2 = 4.0$ m, $h_4 = h_5 = h_6 = h_7 = 1.0$ m < 0.4 B

$\gamma_0 = \gamma_1 = 1.8$ (ton/m³) $= 17.7$ (kN/m³), $E_{s_1} = 40,000$ (kPa), $\beta_1 = 0.76$

$\gamma_2 = 2.0$ (ton/m³), $E_{s_2} = 25,000$ (kPa), $\beta_2 = 0.72$

Engineering properties of soils are determined by field and laboratory methods.

Required. Compute settlement of soil under footing.

Solution. $\sigma_p = \dfrac{P_v}{B \cdot L} = \dfrac{3000}{3 \times 5.4} = 185.2$ kPa, $\sigma_{\gamma_0} = \gamma_0 h_0 = 17.7 \times 2.0 = 35.4$ kPa

$\sigma_{a_0} = \sigma_p - \sigma_{\gamma_0} = 185.2 - 35.4 = 149.8$ kPa, $0.2\sigma_\gamma = 0.2 \times \gamma_{1(2)}(h_0 + z_i)$(kPa)

$\sigma_{a_i} = \alpha_i \times \sigma_{a_0}$ (for α_i see Table 8.5a), $L/B = 5.4/3.0 = 1.8$

H_i	$z_i(m)$	z_i/B	α_i	σ_{a_i}(kPa)	$0.2\sigma_\gamma$ (kPa)
H_1	$z_1 = 0.5$	0.167	0.944	141.4	8.9
	$z_2 = 1.5$	0.500	0.794	118.9	12.4
	$z_3 = 2.5$	0.833	0.561	84.0	15.9
H_2	$z_4 = 3.5$	1.167	0.391	58.4	21.6
	$z_5 = 4.5$	1.500	0.282	42.2	25.5
	$z_6 = 5.5$	1.833	0.207	31.0	29.6
	$z_7 = 6.5$	2.167	0.157	23.5	33.3

Assume. $z = 6.0$ m, $z/B = 2.0$, $\alpha = 0.189$,

$\sigma_a = 0.189 \times 149.8 = 28.3 \approx 0.2\sigma_\gamma = 0.2(5.0 \times 17.7 + 3.0 \times 19.6) = 29.5$ kPa

Settlement. $S = 1.0(141.4 + 118.8 + 84.0)\dfrac{0.76}{40,000} + 1.0(58.4 + 42.2 + 31.0)\dfrac{0.72}{25,000}$

$= 0.0065 + 0.0038 = 0.0103$ m

TABLE 8.4 Soils: Settlement

Method Based on Elastic Theory

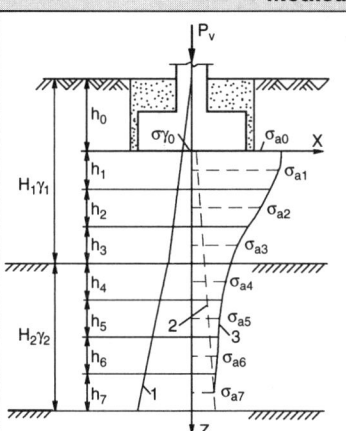

$1 = \text{Line } \sigma_\gamma,\ 2 = \text{Line } 0.2\sigma_\gamma,$
$3 = \text{Line } \sigma_a$

Settlement: $\displaystyle S = \sum_{i=1}^{i=n} \sigma_{a_i} h_i \frac{\beta_i}{E_{s_i}}$

where

n = number of h height layers, $h \le 0.4B$

σ_{a_i} = additional vertical pressure at midheight of h_i layer, $\alpha_{a_i} = \alpha_i \cdot \sigma_{a_0}$

$\sigma_{a_0} = \sigma_P - \sigma_{\gamma_0},\ \sigma_{\gamma_0} = \gamma_0 h_0,\ \sigma_P = \dfrac{P_v}{B \cdot L}$

α_i = coefficient from Table 8.5a
γ_i = unit weight of soil
P_v = total vertical load, $B < L$
B = width of footing, L = length of footing
E_{s_i} = modulus of deformation of soil

$\beta = 1 - \dfrac{2\mu^2}{1-\mu},\ \mu$ = Poisson's ratio for soil

Sand: $\beta = 0.76$, sandy loam: $\beta = 0.72$
Sandy clay: $\beta = 0.57$, clay: $\beta = 0.4$

Alternative Formulas

Settlement of loads on clay due to primary consolidation:	$S = \dfrac{e_0 - e}{1 + e_0}[H]$ e_0 = initial void ratio of soil in situ e = void ratio of soil corresponding to total pressure acting at midheight of consolidating clay layer H = thickness of consolidating clay layer
Settlement of loads on clay due to secondary consolidation:	$S_s = C_\alpha H \cdot \log(t_s/t_p),\ C_\alpha \approx 0.01 - 0.03$ t_s = life of structure or time for which settlement is required t_p = time to completion of primary consolidation

NOTES

Lateral Pressure of Cohesionless Soils

For walls that retain cohesionless soils and are free to move an appreciable amount, the total thrust from the soil is

$$P = \frac{1}{2}\gamma H^2 \cos i \, \frac{\cos i - \sqrt{(\cos i)^2 - (\cos \phi)^2}}{\cos i + \sqrt{(\cos i)^2 - (\cos \phi)^2}}$$

When the surface behind the wall is level, the thrust is

$$P = \tfrac{1}{2}\gamma H^2 K_A$$

where

$$K_A = \left[\tan\left(45° - \frac{\phi}{2}\right)\right]^2$$

The thrust is applied at a point $H/3$ above the bottom of the wall, and the pressure distribution is triangular, with the maximum pressure of $2P/H$ occurring at the bottom of the wall.

For walls that retain cohesionless soils and are free to move only a slight amount, the total thrust is $1.12P$, where P is as given earlier. The thrust is applied as the midpoint of the wall, and the pressure distribution is trapezoidal, with the maximum pressure of $1.4P/H$ extending over the middle six-tenths of the height of the wall.

For walls that retain cohesionless soils and are completely restrained (very rare), the total thrust from the soil is*

$$P = \tfrac{1}{2}\gamma H^2 \cos i \, \frac{\cos i + \sqrt{(\cos i)^2 - (\cos \phi)^2}}{\cos i - \sqrt{(\cos i)^2 - (\cos \phi)^2}}$$

When the surface behind the wall is level, the thrust is

$$P = \tfrac{1}{2}\gamma H^2 K_P$$

where

$$K_P = \left[\tan\left(45° - \frac{\phi}{2}\right)\right]^2$$

The thrust is applied at a point $H/3$ above the bottom of the wall, and the pressure distribution is triangular, with the maximum pressure of $2P/H$ occurring at the bottom of the wall.

Lateral Pressure of Cohesive Soils

For walls that retain cohesive soils and are free to move a considerable amount over a long time, the total thrust from the soil (assuming a level surface) is

$$P = \tfrac{1}{2}\gamma H^2 K_A - 2cH\sqrt{K_A}$$

---

*See page 156 for symbols.

Continued on page 162

Soils

TABLE 8.5 Soils: Settlement

	Coefficient α_I											
	L/B											For
Z_I/B	1.0	1.2	1.4	1.6	1.8	2.0	2.4	2.8	3.2	4.0	5.0	≥10	Circle
0	1.000	1.000	1.000	1.000	1.000	1.000	1.000	1.000	1.000	1.000	1.000	1.000	1.000
0.4	0.800	0.830	0.848	0.859	0.866	0.870	0.875	0.878	0.879	0.880	0.881	0.881	0.756
0.8	0.449	0.496	0.532	0.558	0.578	0.593	0.612	0.623	0.630	0.636	0.639	0.642	0.390
1.2	0.257	0.294	0.325	0.352	0.374	0.392	0.419	0.437	0.469	0.462	0.470	0.477	0.214
1.6	0.160	0.187	0.210	0.232	0.251	0.267	0.294	0.314	0.329	0.348	0.360	0.374	0.130
2.0	0.108	0.127	0.145	0.161	0.176	0.189	0.214	0.233	0.241	0.270	0.285	0.304	0.087
2.4	0.077	0.092	0.105	0.118	0.130	0.141	0.161	0.178	0.192	0.213	0.230	0.258	0.062
2.8	0.058	0.069	0.079	0.089	0.099	0.108	0.124	0.139	0.152	0.172	0.189	0.228	0.046
3.2	0.045	0.053	0.062	0.070	0.077	0.085	0.098	0.110	0.122	0.141	0.158	0.190	0.036
3.6	0.036	0.042	0.049	0.056	0.062	0.068	0.080	0.090	0.100	0.117	0.133	0.175	0.030
4.0	0.029	0.035	0.040	0.046	0.051	0.056	0.066	0.075	0.084	0.095	0.113	0.158	0.025
4.4	0.024	0.029	0.034	0.038	0.042	0.047	0.055	0.063	0.070	0.084	0.098	0.144	0.021
4.8	0.020	0.024	0.028	0.032	0.036	0.040	0.047	0.054	0.060	0.072	0.085	0.132	0.018
5.0	0.019	0.022	0.026	0.030	0.033	0.037	0.044	0.050	0.056	0.067	0.079	0.126	0.017

Method Based on Winkler's Hypothesis

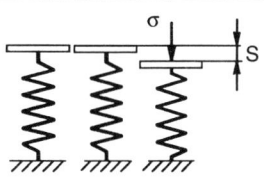

Winkler's support model

Settlement: $S = \dfrac{\sigma}{k_w}$

where σ = compressive stress applied to a unit area of a soil subgrade

S = settlement of unit area of a soil subgrade

k_w = Winkler's coefficient of subgrade reaction (force per length cubed)

$$k_w = \frac{E_s}{h}$$

$$k_{w_{1,2}} = \frac{k_{w_1} \cdot k_{w_2}}{k_{w_1} + k_{w_2}}$$

$$k_{w_1} = \frac{E_{s_1}}{h_1}, \quad k_{w_2} = \frac{E_{s_2}}{h_2}$$

NOTES

or, because highly cohesive soils generally have small angles of internal friction,

$$P = \tfrac{1}{2}\gamma H^2 - 2cH$$

The thrust is applied at a point somewhat below $H/3$ from the bottom of the wall, and the pressure distribution is approximately triangular.

For walls that retain cohesive soils and are free to move only a small amount or not at all, the total thrust from the soil is

$$P = \tfrac{1}{2}\gamma H^2 K_P$$

because the cohesion would be lost through plastic flow.

For slope stability analysis, it is necessary to compute the factor of safety for two or three possible failure surfaces with different diameters. The smallest of the obtained values is then accepted as the result.

Soils

TABLE 8.6 Soils

Modulus of Deformation (E_s) and Winkler's Coefficient (k_w) for Some Types of Soil		
Soil Type	Range (E_s) (MPa)	Range (k_w) (N/cm³)
Crushed stone, gravel sand	55–65	90–150
Coarse-grained sand	40–45	75–120
Medium-grained sand	35–40	60–90
Fine-grained sand	25–35	45–75
Sandy loam	15–25	30–60
Sandy clay	10–30	30–45
Clay	15–30	25–45

Shear Strength of Soil

Coulomb equation: $\tau_s = c + \sigma \tan \phi$

where τ_s = shear strength
c = cohesion
σ = effective intergranular normal pressure
ϕ = angle of internal friction
$\tan \phi$ = coefficient of friction

Slope Stability Analysis

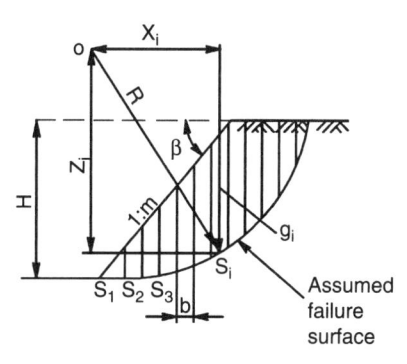

Factor of safety for slope F.S. ≥ 1.5 to 1.8

$$F.S. = \frac{\sum_{i=1}^{i=n} g_i z_i \tan \phi_i + R \sum_{i=1}^{i=n} c_i s_i}{\sum_{i=1}^{i=n} g_i x_i}$$

where g_i = weight of mass for element i
c_i = cohesion of soil
ϕ_i = angle of internal friction
H = depth of cut

Safety depth of cut $\quad H_s = \dfrac{2c}{\gamma} \cdot \dfrac{\cos \phi}{1 - \sin \phi}$

NOTES

Example for Table 8.4. Bearing capacity analysis

Given. Rectangular footing, $B = 3.6$ m, $L = 2.8$ m, $B/L = 1.28$, smooth base
Granular soil, $\phi = 30°$, $c = 0$, $\gamma = 130$ lb/ft^3 = $130 \times 0.1571 = 20.42$ kN/m^3
Loads $P = 2500$ kN, $M = 500$ kN·m, $e = 500/2500 = 0.2$ m,
$e/B = 0.2/3.6 = 0.06$
Bearing capacitiy factors $R_e = 0.78$, $N_q = 20.1$, $N_\gamma = 20$

Required. Compute factor of safety for footing

Solution. $q_{ult} = \gamma D_f N_q + 0.4 \gamma B N_\gamma = 20.42 \times 2 \times 20.1 + 0.4 \times 20.42 \times 3.6 \times 20 = 1409$ kN/m^2
F.S. $= q_{ult} \cdot B \cdot L \cdot R_e / P = 1409 \times 3.6 \times 2.8 \times 0.78 / 2500 = 4.43 > 3$

TABLE 8.7 Bearing Capacity Analysis

Ultimate bearing capacity
Continuous footing (width B):
$$q_{ult} = cN_c + \gamma D_f N_q + 0.5\gamma BN_\gamma$$
Square and rectangular footing (width B, length L):
$$q_{ult} = cN_c\left(1 + 0.3\frac{B}{L}\right) + \gamma D_f N_q + 0.4\gamma BN_\gamma$$

Circular footing (radius R):
$$q_{ult} = 1.3cN_c + \gamma D_f N_q + 0.6\gamma BN_\gamma$$

where c = cohesion of soil
 γ = unit weight of soil
 N_c, N_q, N_γ = Terzaghi's bearing capacity factors
 D_f = distance from ground surface to base of footing

Factor of safety for footing F.S. $\geq$ 2.5 to 3
Continuous footing: F.S. $= q_{ult} \cdot B \cdot R_e / P$
Square and rectangular footing:
  F.S. $= q_{ult} \cdot B \cdot L \cdot R_e / P$
Circular footing: F.S. $= q_{ult} \cdot \pi \cdot R^2 \cdot R_e / P$
where R_e = eccentric load reduction factor

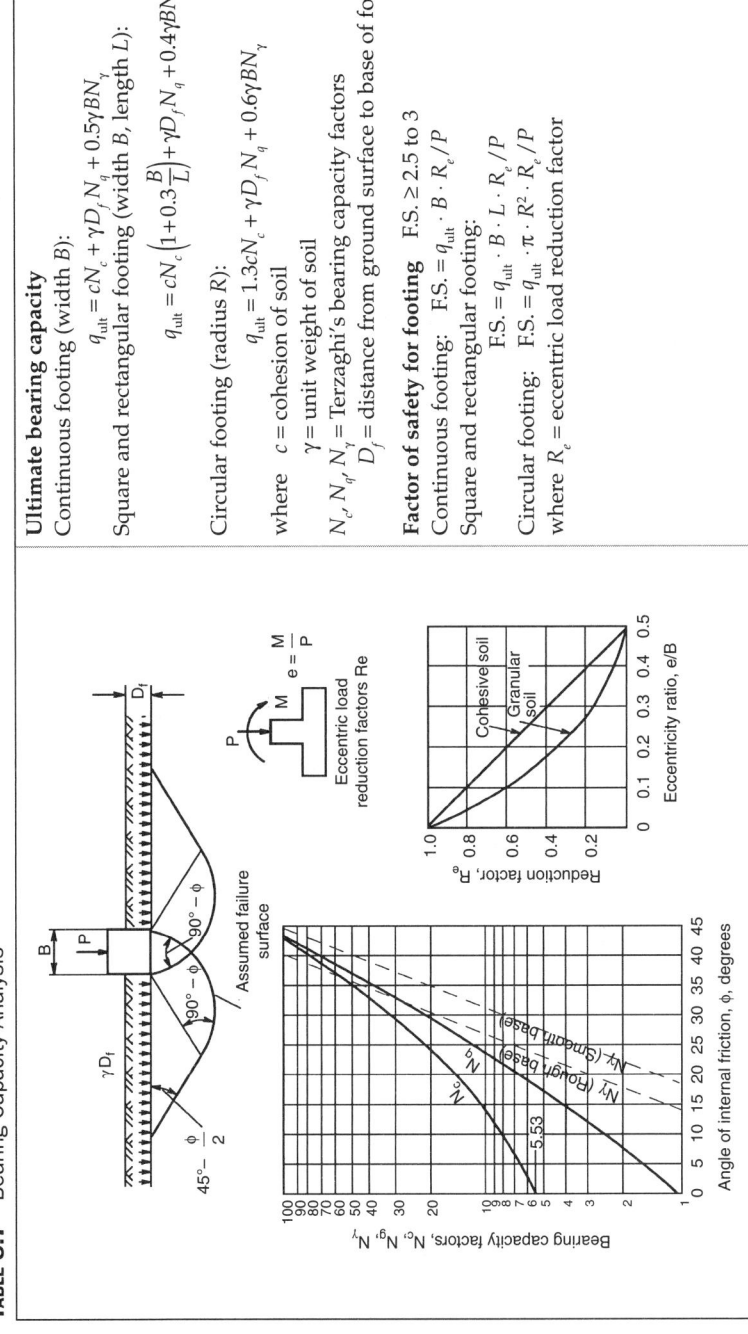

NOTES

Allowable Bearing Pressures

Approximate allowable soil bearing pressures, without tests, for various soil and rocks are given in Table 8.1N for normal conditions. These basic bearing pressures may be increased when the base of the footing is embedded beyond normal depth. Rock values may be increased by 10% for each foot of embedment beyond 4 ft in fully confined conditions, but the values may not exceed twice the basic values.

In any case, bearing pressures should be limited to values such that the proposed construction will be safe against failure of the soil under 100% overload.

TABLE 8.1N Allowable Bearing Pressures for Soils

Soil Material	Pressure, tons/ft^2	Notes
Unweathered sound rock	60	No adverse seam structure
Medium rock	40	
Intermediate rock	20	
Weathered, seamy, or porous rock	2–8	
Hardpan	12	Well cemented
Hardpan	8	Poorly cemented
Gravel soils	10	Compact, well graded
Gravel soils	8	Compact with more than 10% gravel
Gravel soils	6	Loose, poorly graded
Gravel soils	4	Loose, mostly sand
Sand solis	3–6	Dense
Fine sand	2–4	Dense
Clay soils	5	Hard
Clay soils	2	Medium stiff
Silt soils	3	Dense
Silt soils	1½	Medium dense
Compacted fills		Compacted to 90 to 95% of maximum density (ASTM D1557)
Fills and soft soils	2–4	By field or laboratory test only

CHAPTER 9
Foundations

NOTES

Tables 9.1 to 9.7 consider two cases of foundation analysis.

1. The footing is supported directly by the soil: Maximum soil reaction (contact pressure) is determined and compared with requirements of the norms or the results of laboratory or field soil research.
2. The footing is supported by the piles: Forces acting on the piles are computed and compared with the pile capacity provided in the catalogs.

If necessary, pile capacity can be computed using the formulas provided in Table 9.4.

TABLE 9.1 Foundations: Direct Foundations

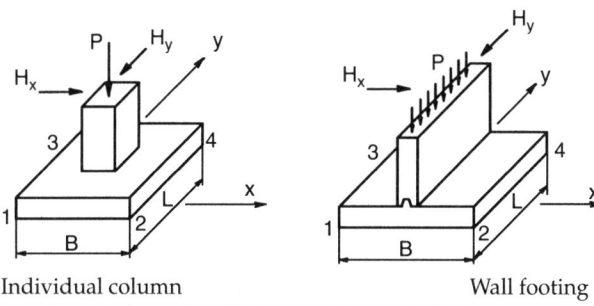

| Individual column | Wall footing |

Contact Pressure and Soil Pressure Diagrams

Two-way action: $q_i = \dfrac{P_v}{A} \pm \dfrac{M_x}{S_x} \pm \dfrac{M_y}{S_y}$ where $A = B \cdot L$, $S_x = \dfrac{B \cdot L^2}{6}$, $S_y = \dfrac{B^2 \cdot L}{6}$.

One-Way Action

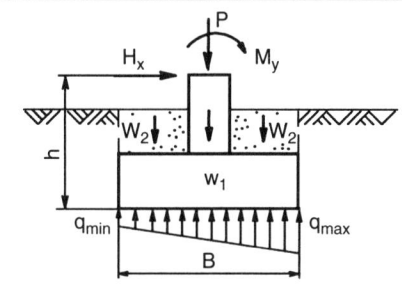

$q_{max} = \dfrac{P_v}{A} + \dfrac{\sum M_y}{S_y}$, $q_{min} = \dfrac{P_v}{A} - \dfrac{\sum M_y}{S_y}$

where $P_v = P + W_1 + 2W_2$

$\sum M_y = H_x \cdot h + M_y$

P = load on footing from column
W_1 = weight of concrete, including pedestal and base pad
W_2 = weight of soil

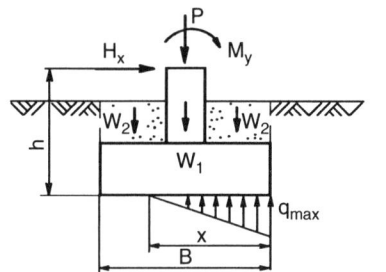

If $q_{min} < 0$, assume $q_{min} = 0$

(soil cannot furnish any tensile resistance)

$x = \dfrac{3\left(P_v \cdot B - 2\sum M_y\right)}{2P_v}$

$q_{max} = \dfrac{2P_v}{x \cdot L}$

NOTES

Example for Tables 9.1 and 9.2. Direct foundation in Table 9.1

Given. Reinforced concrete footing, $B = 3.6$ m, $L = 2.8$ m, $h = 3$ m
$A = B \cdot L = 3.6 \times 2.8 = 10.08$ m², $S_y = L \cdot B^2 / 6 = 6.048$ m³
Loads $P_v = P + W_1 + 2W_2 = 2250$ kN, $M_y = 225$ kN·m, $H = 200$ kN
Allowable soil contact pressure $\sigma = 360$ kPa $= 360$ kN/m², $f = 0.4$

Required. Compute contact pressure, factors of safety against sliding and overturning.

Solution. $q_{max} = \dfrac{P_v}{A} + \dfrac{\sum M_y}{S_y}$, $q_{min} = \dfrac{P_v}{A} - \dfrac{\sum M_y}{S_y}$

$q_{max} = \dfrac{2250}{10.08} + \dfrac{200 \times 3 + 225}{6.048} = 223.2 + 136.4 = 359.6 < 360$ kPa

$q_{min} = 223.2 - 136.4 = 86.8$ kPa

Factor of safety against sliding $\text{F.S.} = \dfrac{P_v \cdot f}{\sum H} = \dfrac{2250 \times 0.4}{200} = 4.5$

Factor of safety against overturning $\text{F.S.} = \dfrac{M_{r(k)}}{M_{0(k)}} = \dfrac{P_v \cdot B/2}{M + \sum H \cdot h}$

$= \dfrac{2250 \times 3.6/2}{225 + 200 \times 3} = 4.9$

Allowable Loads on Piles

A dynamic formula extensively used in the United States to determine the allowable static load on a pile is the *Engineering News* formula. For piles driven by a drop hammer, the allowable load is

$$P_a = \dfrac{2WH}{p+1} \tag{9.1}$$

For piles driven by a steam hammer, the allowable load is

$$P_a = \dfrac{2WH}{p+0.1} \tag{9.2}$$

where P_a = allowable pile load, tons (kg)
W = weight of hammer, tons (kg)
H = height of drop, ft (m)
p = penetration of pile per blow, in (mm)

The preceding two equations include a factor of safety of 6.

Foundations

TABLE 9.2 Foundations

Direct Foundation Stability	
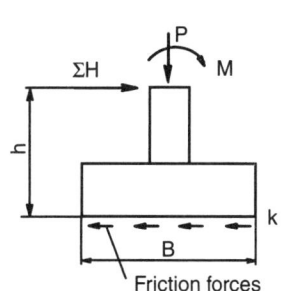	Factor of safety against sliding: $\text{F.S.} = \dfrac{P_v \cdot f}{\sum H}$ P_v = total vertical load, $\sum H$ = total horizontal forces f = coefficient of friction between base and soil $f \approx 0.4 - 0.5$
	Factor of safety against overturning: $\text{F.S.} = \dfrac{M_{r(k)}}{M_{o(k)}}$ $M_{r(k)} = P_v \cdot B/2$, $\quad M_{o(k)} = M + \sum H \cdot h$ $M_{r(k)}$ = moment to resist turning $M_{o(k)}$ = turning moment
Pile Foundations	
Distribution of Loads in Pile Group	
Example 9.2a 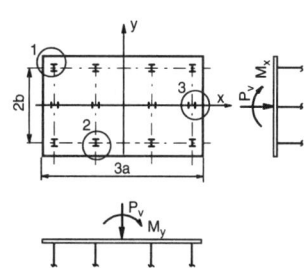 Foundation plan and sections	Axial load on any particular pile: $P_i = \dfrac{P_v}{n \cdot m} \pm \dfrac{M_y \cdot x}{\sum (x)^2} \pm \dfrac{M_x \cdot y}{\sum (y)^2}$ P_v = total vertical load acting on pile group n = number of piles in a row m = number of rows of pile M_x, M_y = moment with respect to x and y axes, respectively x, y = distance from pile to y and x axes, respectively Example 9.2a: $n = 4$, $m = 3$ $\sum (x)^2 = 2 \cdot 3[(0.5a)^2 + (1.5a)^2] = 6 \cdot 6.25a = 13.5a$ $\sum (y)^2 = 2 \cdot 4 \cdot (b)^2 = 8b^2$
Pile 1: $x = -1.5a$, $y = b$,	$P_1 = \dfrac{P_v}{4 \cdot 3} - \dfrac{M_y \cdot 1.5a}{13.5a^2} + \dfrac{M_x \cdot b}{8b^2} = \dfrac{P_v}{12} - \dfrac{M_y}{9a} + \dfrac{M_x}{8b}$
Pile 2: $x = -0.5a$, $y = -b$,	$P_2 = \dfrac{P_v}{4 \cdot 3} - \dfrac{M_y \cdot 0.5a}{13.5a^2} - \dfrac{M_x \cdot b}{8b^2} = \dfrac{P_v}{12} - \dfrac{M_y}{27a} - \dfrac{M_x}{8b}$
Pile 3: $x = 0.5a$, $y = 0$,	$P_3 = \dfrac{P_v}{4 \cdot 3} + \dfrac{M_y \cdot 1.5a}{13.5a^2} + \dfrac{M_x \cdot 0}{8b^2} = \dfrac{P_v}{12} + \dfrac{M_y}{9a}$

Toe Capacity Load on Piles

For piles installed in cohesive soils, the ultimate tip load may be computed from

$$Q_{bu} = A_b q = A_b N_c c_u \tag{9.3}$$

where A_b = end-bearing area of pile, ft² (m²)
q = bearing capacity of soil, tons/ft² (MPa)
N_c = bearing-capacity factor
c_u = undrained shear strength of soil within zone 1 pile diameter above and 2 diameters below pile tip, psi (MPa)

Although theoretical conditions suggest that N_c may vary between about 8 and 12, N_c is usually taken as 9.

For cohesionless soils, the toe resistance stress q is conventionally expressed by Eq. (9.1) in terms of a bearing-capacity factor N_q and the effective overburden pressure at the pile tip σ'_{vo}.

$$q = N_q \sigma'_{vo} \leq q_l \tag{9.4}$$

Some research indicates that, for piles in sands, q, like f_s, reaches a quasi-constant value q_l, after penetrations of the bearing stratum in the range of 10 to 20 pile diameters. Approximately

$$q_l = 0.5 N_q \tan \phi \tag{9.5}$$

where ϕ is the friction angle of the bearing soils below the critical depth. Values of N_q applicable to piles are given in Fig. 9.1. Empirical correlations of soil test data with q and q_l have also been applied to predict successfully end-bearing capacity of piles in sand.

Foundations, Substructures, and Superstructures

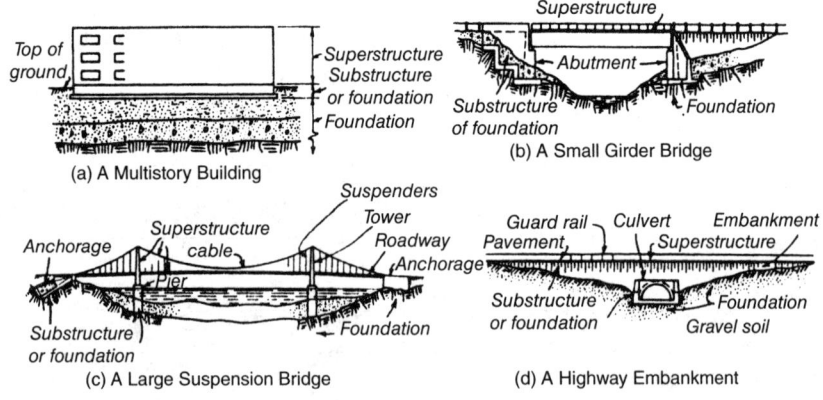

FIGURE 9.1 Foundations, substructures, and superstructures in various types of structures.

TABLE 9.3 Foundations

Distribution of Loads in Pile Group

Example 9.2b

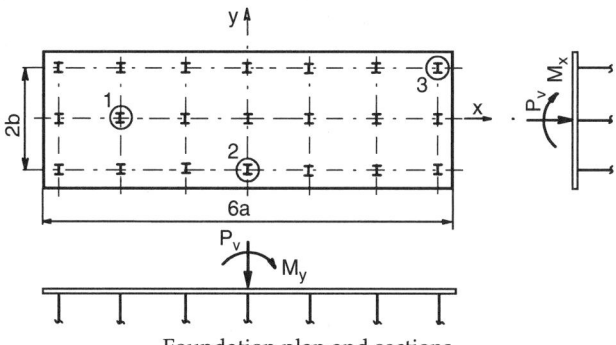

Foundation plan and sections

Axial load on any particular pile: $P_i = \dfrac{P_v}{n \cdot m} \pm \dfrac{M_y \cdot x}{\sum (x)^2} \pm \dfrac{M_x \cdot y}{\sum (y)^2}$

$n = 7, \quad m = 3, \quad \sum (x)^2 = 2 \cdot 3 \cdot [(a)^2 + (2a)^2 + (3a)^2] = 6 \cdot 14a^2 = 84a^2$

$\sum (y)^2 = 2 \cdot 7 \cdot (b)^2 = 14b^2$

Pile 1: $x = -2a, \quad y = 0, \quad P_1 = \dfrac{P_v}{7 \cdot 3} - \dfrac{M_y \cdot 2a}{84a^2} + \dfrac{M_x \cdot 0}{14b^2} = \dfrac{P_v}{21} - \dfrac{M_y}{42a}$

Pile 2: $x = 0, \quad y = -b, \quad P_2 = \dfrac{P_v}{7 \cdot 3} + \dfrac{M_y \cdot 0}{84a^2} - \dfrac{M_x \cdot b}{14b^2} = \dfrac{P_v}{21} - \dfrac{M_x}{14b}$

Pile 3: $x = 3a, \quad y = b, \quad P_3 = \dfrac{P_v}{7 \cdot 3} + \dfrac{M_y \cdot 3a}{84a^2} + \dfrac{M_x \cdot b}{14b^2} = \dfrac{P_v}{21} + \dfrac{M_y}{28a} + \dfrac{M_x}{14b}$

Maximum and minimum axial load on pile:

$P_{\substack{\max \\ \min}} = \dfrac{P_v}{n \cdot m} \pm \dfrac{M_y}{S_x} \pm \dfrac{M_x}{S_y}, \quad S_x = \dfrac{n(n+1)a \cdot m}{6}, \quad S_y = \dfrac{m(m+1)b \cdot n}{6}$

In Example 9.2b: $S_x = \dfrac{7(7+1)a \cdot 3}{6} = 28a, \quad S_y = \dfrac{3(3+1)b \cdot 7}{6} = 14b$

Pile Group Capacity

$N_g = E_g \cdot n \cdot m \cdot N_p$

Converse-Labarre equation:

$E_g = 1 - \left(\dfrac{\theta}{90}\right) \dfrac{(n-1)m + (m-1)n}{n \cdot m}$

For cohesionless soil $E_g = 1.0$

Where N_g = capacity of pile group
E_g = pile group efficiency
N_p = capacity of single pile
$\theta = \arctan d/s$ (deg), d = diameter of piles,
s = min. spacing of piles, center to center

Determining Foundation Settlement from Soil Test Borings

When an important structure is being planned, test borings of the soil under it can give important data on possible settlement of the foundation. Results of two test borings at loads of 1 and 2 tons/sq. ft are as follows:

Boring no. 1, Sample no. 2: Stiff Blue Clay	
Unit Pressure p, lb/ft²	Voids Ratio e
0	0.886
2000	0.864
4000	0.849

These data may serve to plot a curve of unit pressure vs. settlement by using the equation

$$\Delta h = \frac{e_i - e_p}{1 + e_i} h \tag{9.6}$$

where Δh = total expected settlement under load, in; e_i = initial voids ratio; e_p = voids ratio when under a pressure p; and h = thickness of layer, in. Notice that e_i is to be the voids ratio of the soil under the pressure estimated as existing at the location.

Using the data obtained from a series a borings, plot the chart shown in Fig. 9.2. This is based on a 2:1 slope distribution of the load on the soil. This slope is often assumed in early calculations. Solving for the two samples shown gives

$$\Delta h_{2000} = \left(\frac{0.886 - 0.864}{1 + 0.886}\right) 10 \times 12 = 1.4 \text{ in}$$

$$\Delta h_{4000} = \left(\frac{0.886 - 0.849}{1 + 0.886}\right) 10 \times 12 = 2.35 \text{ in}$$

Additional tests provide the data shown in Fig. 9.2. This chart will be helpful in applying typical settlement calculations.

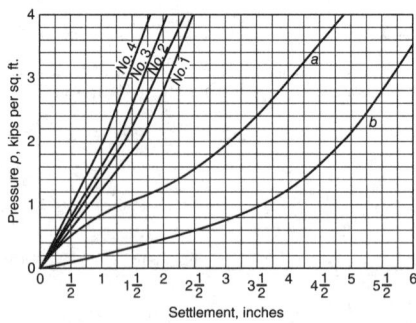

Figure 9.2 Load-settlement curves for a 10-ft (3-m) stratum of soil, based on lab tests of soil.

Continued on page 176

TABLE 9.4 Foundations

Pile capacity

$$Q_u = Q_{fr} + Q_{tip}$$

where Q_u = ultimate (at failure) bearing capacity of a single pile
Q_{fr} = bearing capacity furnished by friction between the soil and the sides of pile
Q_{tip} = bearing capacity furnished by the soil just below the pile's tip

$$Q_{fr} = f_s \cdot C_p \left[0.5\gamma \cdot D_c^2 + \gamma \cdot D_c(H - D_c) \right] \cdot K, \quad Q_{tip} = \gamma \cdot D_c \cdot N_q \cdot A_{tip}$$

where f_s = coefficient of friction between soil and pile. Concrete: f_s = 0.45, wood: f_s = 0.4, steel: f_s = 0.2 + 0.4
C_p = circumference of pile
γ = unit weight of soil
D_c = critical depth, ranging approximately from 10 pile diameters for loose sand to 20 pile diameters for dense, compact sand
H = embedded length of pile
K = coefficient of lateral soil pressure
N_q = bearing capacity factor (see Table 8.7)
A_{tip} = area of the pile tip

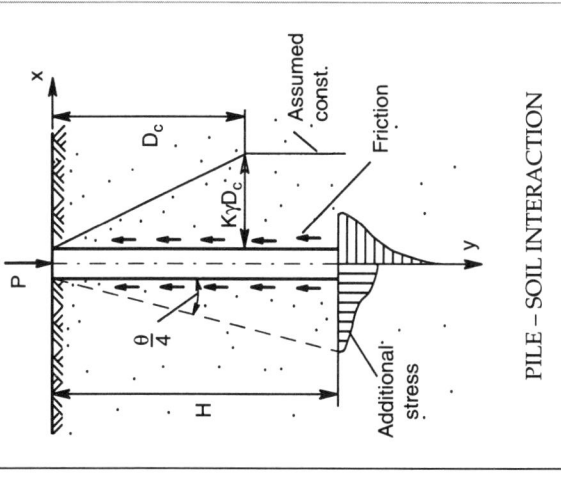

PILE – SOIL INTERACTION

θ = angle of internal friction

NOTES

Now compare the effects of a large loaded area and a small one upon the unit pressures assumed to exist at a plane in a stratum below the structures. Figure 9.3a pictures a footing AB and large structure GH. Assume that the intensity of pressure at AB and GH equals 4000 lb/ft³. According to the assumption of 2:1 distribution, the intensities of pressures at 20- and 40-ft depths below these structures are the following:

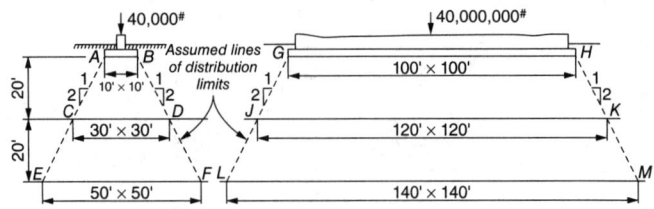

(a)-Comparison of relative distribution of pressure under small and large foundations

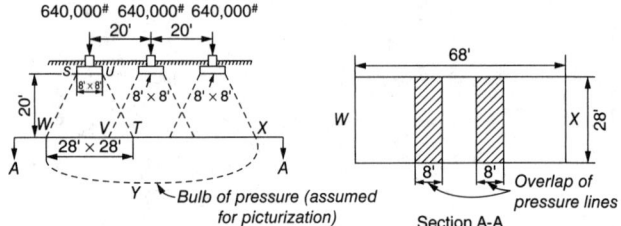

(b)-Distribution of pressure from a row of footings

FIGURE 9.3 A study of the distribution of pressure in soils under various foundations, assuming 2:1 distribution.

$$\text{At } CD, p_{20} = \frac{400,000}{30 \times 30} = 445 \text{ lb/ft}^2$$

$$\text{At } EF, p_{40} = \frac{400,000}{50 \times 50} = 160 \text{ lb/ft}^2$$

$$\text{At } JK, p_{20} = \frac{40,000,000}{120 \times 120} = 2780 \text{ lb/ft}^2$$

$$\text{At } LM, p_{40} = \frac{40,000,000}{140 \times 140} = 2040 \text{ lb/ft}^2$$

The Boussinesq equation

$$\sigma_3 = \frac{3Q}{2\pi z^2}\left[\frac{1}{1+(r/z)^2}\right]^{5/2} \tag{9.7}$$

may be used to estimate the vertical pressure at some point at depth z and horizontal offset r caused by a concentrated load Q. This formula assumes that the earth mass is homogeneous, elastic, and isotropic, which it seldom ever is. On the other hand, it gives a reasonable idea as to the probable magnitude of the vertical pressure, and this information is very useful.

Foundations

TABLE 9.5 Foundations: Rigid Continuous Beam Elastically Supported

The following method can be applied on condition that $L \leq 0.8 \cdot h \cdot \sqrt[3]{E/E_s}$ where E, L, and h are the modulus of elasticity, length of beam, and depth of beam, respectively.
E_s = modulus of deformation of soil

	Uniformly Distributed Load (w)					
1	Soil Reaction: $q_i = \alpha_{q(i)} \cdot w$					
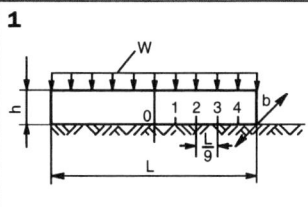 Soil reaction diagram	b/L	$\alpha_{q(0)}$	$\alpha_{q(1)}$	$\alpha_{q(2)}$	$\alpha_{q(3)}$	$\alpha_{q(4)}$
	0.33	0.799	0.832	0.858	0.907	1.494
	0.22	0.846	0.855	0.881	0.927	1.408
	0.11	0.889	0.890	0.919	0.961	1.298
	0.07	0.900	0.905	0.928	0.973	1.247
	Bending Moment: $M_i = \alpha_{m(i)} \cdot w \cdot b \cdot L^2$					
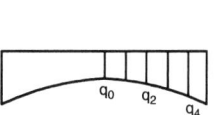 Moment diagram	b/L	$\alpha_{m(0)}$	$\alpha_{m(1)}$	$\alpha_{m(2)}$	$\alpha_{m(3)}$	$\alpha_{m(4)}$
	0.33	0.018	0.014	0.010	0.006	0.001
	0.22	0.012	0.011	0.009	0.005	0.001
	0.11	0.009	0.008	0.006	0.004	0.000
	0.07	0.008	0.007	0.006	0.003	0.000
	Shear: $V_i = \alpha_{v(i)} \cdot w \cdot b \cdot L$					
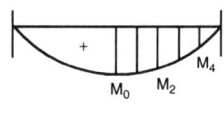 Shear diagram	b/L	$\alpha_{v(0)}$	$\alpha_{v(1)}$	$\alpha_{v(2)}$	$\alpha_{v(3)}$	$\alpha_{v(5)}$
	0.33	0.0	−0.019	−0.037	−0.050	−0.027
	0.22	0.0	−0.016	−0.030	−0.041	−0.023
	0.11	0.0	−0.014	−0.024	−0.031	−0.016
	0.07	0.0	−0.012	−0.020	−0.026	−0.014

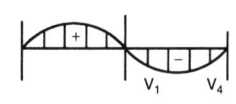

NOTES

Estimates of Settlement

1. *Settlement of a structure on a thick plastic soil.* Assume that the structure shown in Fig. 9.4a is to have a heavy concrete mat at its base, and that the unit pressure at AB is to be 2 tons/ft². The soil is a 40-ft layer of clay above a thick stratum of sand. Assume further that undisturbed soil samples were taken at 10-ft intervals, as indicated, and that the pressure-settlement curves for a 10-ft depth as determined by laboratory tests are as pictured in Fig. 9.2. How much settlement may be expected?

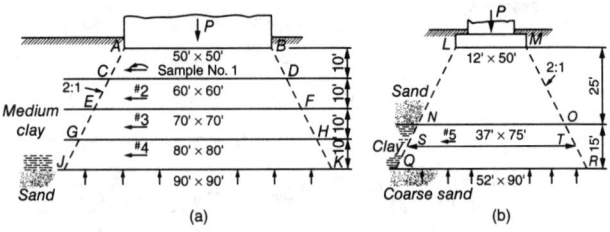

FIGURE 9.4 Procedure for roughly estimating settlements.

For convenience, the 2:1 distribution will be assumed. The estimate will be made in three ways for purposes of comparison.

a. Use of average of unit pressures at top and bottom of stratum, found as follows:

$$P = 50 \times 50 \times 4 = 10{,}000 \text{ kips}$$
$$p_{AB} = 4 \text{ kips/ft}^2$$
$$p_{JK} = 10{,}000/90^2 = 1.2 \text{ kips/ft}^2$$
$$\text{Average } p = \frac{4 + 1.2}{2} = 2.6 \text{ kips/ft}^2$$

Average settlement Δh for 10 ft, interpolated for $p = 2.6$ from the mean of the curves for samples no. 2 and 3 of Fig. 9.2 is 1⅝ in. Therefore, the approximate total $\Delta h = 4 \times 1⅝ = 6½$ in.

b. Use of unit pressure at middle of stratum:

$$p_{EF} = 10{,}000/70^2 = 2 \text{ kips/ft}^2$$

Average Δh for 10 ft determined as in (a) is 1⅜ in. Therefore, the approximate total $\Delta h = 4 \times 1⅜ = 5½$ in.

Continued on page 180

TABLE 9.6 Foundations: Rigid Continuous Beam Elastically Supported

	Concentrated Loads					
2	**Bending Moment:** $M_i = \alpha_{m(i)} \cdot P \cdot L$					
	b/L	$\alpha_{m(0)}$	$\alpha_{m(1)}$	$\alpha_{m(2)}$	$\alpha_{m(3)}$	$\alpha_{m(4)}$
	0.33	0.130	0.087	0.048	0.019	0.003
	0.22	0.134	0.085	0.046	0.018	0.003
	0.11	0.131	0.082	0.044	0.017	0.002
	0.07	0.129	0.081	0.043	0.016	0.002
	Shear: $V_i = \alpha_{v(i)} \cdot P$					
	b/L	$\alpha_{v(0)}$	$\alpha_{v(1)}$	$\alpha_{v(2)}$	$\alpha_{v(3)}$	$\alpha_{v(4)}$
	0.33	−0.500	−0.408	−0.314	−0.216	−0.083
	0.22	−0.500	−0.404	−0.308	−0.208	−0.078
	0.11	−0.500	−0.402	−0.302	−0.197	−0.072
	0.07	−0.500	−0.400	−0.298	−0.192	−0.069
3	**Bending Moment:** $M_i = \alpha_{m(i)} \cdot P \cdot L$					
	b/L	$\alpha_{m(0)}$	$\alpha_{m(1)}$	$\alpha_{m(2)}$	$\alpha_{m(3)}$	$\alpha_{m(4)}$
	0.33	0.050	0.063	0.096	0.038	0.006
	0.22	0.046	0.059	0.092	0.036	0.005
	0.11	0.040	0.053	0.088	0.034	0.004
	0.07	0.030	0.051	0.086	0.032	0.003
	Shear: $V_i = \alpha_{v(i)} \cdot P$					
	b/L	$\alpha_{v(0)}$	$\alpha_{v(1)}$	$\alpha_{v(2)}$	$\alpha_{v(3)}$	$\alpha_{v(4)}$
	0.33	+0.184	+0.372	−0.628	−0.432	−0.166
	0.22	+0.191	+0.384	−0.616	−0.416	−0.156
	0.11	+0.196	+0.396	−0.604	−0.395	−0.144
	0.07	+0.201	+0.404	−0.596	−0.385	−0.138

NOTES

c. Use of a series of imaginary 10-ft layers:
In this case the 40-ft depth is divided into 10-ft layers. The computations are tabulated as follows, with values of the settlement for each layer scaled from the proper curve in Fig. 9.2.

Stratum	Sample No.	p, kips/ft^2	Δh, in per 10-ft Depth
AB-CD	1	$10{,}000/55^2 = 3.3$	2⅛
CD-EF	2	$10{,}000/65^2 = 2.4$	1⅝
EF-GH	3	$10{,}000/75^2 = 1.8$	1⅛
GH-JK	4	$10{,}000/85^2 = 1.4$	¾
			$\Sigma \Delta h = 5⅝$

All these results are estimates only.

Example for Table 9.7. Rigid continuous footing 4 in Table 9.7

Given. Reinforced concrete footing, $L = 6$ m, $b = 2$ m, $h = 1$ m, $b/L = 0.33$
$E = 3370$ kips/in$^2 = 3370 \times 6.8948 = 23{,}235$ MPa
$E_s = 40$ MPa, concentrated loads $P = 200$ kN

Required. Compute M_0, M_3, V_3^L, and V_3^R.

Solution. Checking condition: $L \le 0.8 \cdot h \cdot \sqrt[3]{E/E_s}$, $6 < 0.8 \times 1 \times \sqrt[3]{23{,}235/40} = 6.672$

$M_0 = \alpha_{m(0)} \times P \times L = -0.061 \times 200 \times 6 = -73.2$ kN·m

$M_3 = \alpha_{m(3)} \times P \times L = 0.038 \times 200 \times 6 = 45.6$ kN·m

$V_3^L = \alpha_{v(3)} \times P = 0.568 \times 200 = 113.6$ kN

$V_3^R = -0.432 \times 200 = -86.4$ kN

TABLE 9.7 Foundations: Rigid Continuous Beam Elastically Supported

	Concentrated Loads					
4	**Bending Moment:** $M_i = \alpha_{m(i)} \cdot P \cdot L$					
	b/L	$\alpha_{m(0)}$	$\alpha_{m(1)}$	$\alpha_{m(2)}$	$\alpha_{m(3)}$	$\alpha_{m(4)}$
	0.33	−0.061	−0.048	−0.015	+0.038	+0.006
	0.22	−0.065	−0.052	−0.019	+0.036	+0.005
	0.11	−0.071	−0.058	−0.023	+0.034	+0.004
	0.07	−0.075	−0.060	−0.025	+0.032	+0.004
	Shear: $V_i = \alpha^L_{v(i)} \cdot P$					
	b/L	$\alpha_{v(1)}$	$\alpha_{v(2)}$	$\alpha^L_{v(3)}$	$\alpha^R_{v(3)}$	$\alpha_{v(4)}$
	0.33	+0.184	+0.372	+0.568	−0.432	−0.166
	0.22	+0.191	+0.384	+0.584	−0.416	−0.156
	0.11	+0.196	+0.396	+0.605	−0.395	−0.144
	0.07	+0.211	+0.404	+0.615	−0.385	−0.138
5	**Bending Moment:** $M_i = \alpha_{m(i)} \cdot P \cdot L$					
	b/L	$\alpha_{m(0)}$	$\alpha_{m(1)}$	$\alpha_{m(2)}$	$\alpha_{m(3)}$	$\alpha_{m(4)}$
	0.33	−0.172	−0.159	−0.126	−0.073	+0.006
	0.22	−0.176	−0.163	−0.130	−0.075	+0.005
	0.11	−0.182	−0.169	−0.134	−0.077	+0.004
	0.07	−0.186	−0.171	−0.136	−0.079	+0.004
	Shear: $V_i = \alpha_{v(i)} \cdot P$					
	b/L	$\alpha_{v(1)}$	$\alpha_{v(2)}$	$\alpha_{v(3)}$	$\alpha^L_{v(4)}$	$\alpha^R_{v(4)}$
	0.33	+0.184	+0.372	+0.568	+0.834	−0.166
	0.22	+0.191	+0.384	+0.584	+0.844	−0.156
	0.11	+0.196	+0.396	+0.605	+0.856	−0.144
	0.07	+0.201	+0.404	+0.615	+0.862	−0.138

Determination of Foundation Footing Size by Housel's Method

A square foundation footing is to transmit a load of 80 kips (355.8 kN) to a cohesive soil, the settlement being restricted to ⅝ in (15.9 mm). Two test footings were loaded at the site until the settlement reached this value. The results were as follows:

Footing Size	Load, lb (N)
1 ft 6 in × 2 ft (45.72 × 60.96 cm)	14,200 (63,161.6)
3 × 3 ft (91.44 × 91.44 cm)	34,500 (153,456.0)

Applying Housel's method, determine the size of the footing in plan.

Calculation Procedure

1. *Determine the values of p and s corresponding to the allowable settlement.* Housel considers that the ability of a cohesive soil to support a footing stems from two sources: bearing strength and shearing strength. This concept is embodied in

$$W = Ap + Ps \qquad (9.8)$$

where W = total load, A = area of contact surface, P = perimeter of contact surface, p = bearing stress directly below footing, and s = shearing stress along perimeter.

Applying the given data for the test footings gives: footing 1, $A = 3$ ft² (2787 cm²), $P = 7$ ft (2.1 m); footing 2, $A = 9$ ft² (8361 cm²), $P = 12$ ft (3.7 m). Then $3p + 7s = 14{,}200$; $9p + 12s = 34{,}500$; $p = 2630$ lb/ft² (125.9 kPa); $s = 900$ lb/lin. ft (13,134.5 N/m).

2. **Compute the size of the footing to carry the specified load.** Let x denote the side of the footing. Then $2630x^2 + 900(4x) = 80{,}000$; $x = 4.9$ ft (1.5 m). Make the footing 5 ft (1.524 m) square.

PART IV
Retaining Structures, Pipes, and Tunnels

CHAPTER 10
Retaining Structures

CHAPTER 11
Pipes and Tunnels: Bending Moments for Various Static Loading Conditions

CHAPTER 10
Retaining Structures

NOTES

For determining the lateral earth pressure on walls of structures, the methods that have proved most popular in engineering practice are those based on analysis of the sliding prism's standing balance. The magnitude of the lateral earth pressure is dependent on the direction of the wall movement. This correlation is represented graphically in Table 10.1. The three known coordinates on the graph are P_a, P_0, and P_p. As the graph demonstrates, the active pressure is the smallest, and the passive pressure the largest, among the forces and reactions acting between the soil and the wall.

Construction experience shows that even a minor movement of the retaining walls away from the soil in many cases leads to the formation of a sliding prism and produces active lateral pressure.

Retaining Structures

TABLE 10.1 Retaining Structures: Lateral Earth Pressure on Retaining Walls

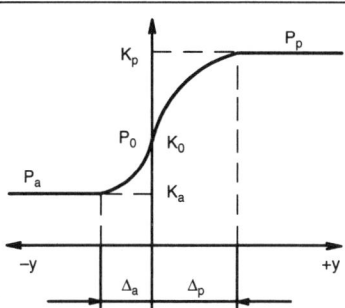

Correlation between lateral earth pressure and wall movement
P_0 = lateral earth pressure at rest
P_a = active lateral earth pressure
P_p = passive lateral earth pressure
K_0, K_a, K_p = coefficients

Coefficients of lateral earth pressure:
K_0 = coefficient of earth pressure at rest:

$$K_0 = \frac{\sigma_h}{\sigma_v} = \frac{\mu}{1-\mu}$$

where σ_h, σ_v = lateral and vertical stresses, respectively
μ = Poisson's ratio

Type of Soil	μ
Sand	0.29
Sandy loam	0.31
Sandy clay	0.37
Clay	0.41

Alternative formulas:
$K_0 = 1 - \sin\phi$ for sands
$K_0 = 0.19 + 0.233 \log(PI)$ for clays
where PI = soil's plasticity index

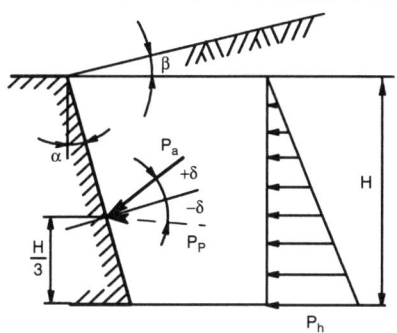

Coulomb earth pressure

$P_a = 0.5 K_a \gamma H^2, \quad P_p = 0.5 K_p \gamma H^2$

where γ = unit weight of backfill soil

K_a = coefficient of active earth pressure
K_p = coefficient of passive earth pressure
Coulomb theory

$$K_a = \frac{\cos^2(\phi - \alpha)}{\left[1 + \sqrt{\frac{\sin(\phi+\delta)\sin(\phi-\beta)}{\cos(\alpha+\delta)\cos(\beta-\alpha)}}\right]^2 \cos^2\alpha \cdot \cos(\alpha+\delta)}$$

$$K_p = \frac{\cos^2(\phi - \alpha)}{\left[1 - \sqrt{\frac{\sin(\phi+\delta)\sin(\phi+\beta)}{\cos(\alpha-\delta)\cos(\beta-\alpha)}}\right]^2 \cos^2\alpha \cdot \cos(\alpha-\delta)}$$

where ϕ = angle of internal friction of backfill soil
δ = angle of friction between wall and soil ($\delta \approx 2/3\phi$)
β = angle between backfill surface line and a horizontal line
α = angle between back side of wall and a vertical line

Earthquake

$$K_{aE} = \frac{\cos^2(\phi - \theta - \alpha)}{\left[1 + \sqrt{\frac{\sin(\phi+\delta)\sin(\phi-\theta-\beta)}{\cos(\alpha+\delta+\theta)\cos(\beta-\alpha)}}\right]^2 \cos\theta \cdot \cos^2\alpha \cdot \cos(\alpha+\theta+\delta)}$$

$\theta = \arctan[K_h/(1-K_v)]$
K_h = seismic coefficient,
$K_h = A_E/2$
A_E = acceleration coefficient
K_v = vertical acceleration coefficient

Retaining Structures, Pipes, and Tunnels

NOTES

Example for Table 10.2. Retaining wall 1 in Table 10.2, $H = 10$ m

Given. Cohesive soil, angle of friction $\phi = 26°$
Cohesion $c = 150$ lb/ft² = $150 \times 47.88 = 7182$ Pa = 7.2 kN/m²
Unit weight of backfill soil $\gamma = 115$ lb/ft³ = $115 \times 0.1571 = 18.1$ kN/m³

Required. Compute active and passive earth pressure per unit length of wall: P_a, h, P_p, d_p.

Solution. Active earth pressure:

$$K_a = \tan^2\left(45° - \frac{\phi}{2}\right) = \tan^2\left(45° - \frac{26°}{2}\right) = 0.39$$

$$p_h = K_a \gamma H - 2c\sqrt{K_a} = 0.39 \times 18.1 \times 10 - 2 \times 7.2\sqrt{0.39} = 61.61 \text{ kN/m}$$

$$h = \frac{p_h H}{p_h + 2c\tan(45° - \phi/2)} = \frac{61.61 \times 10}{61.61 + 2 \times 7.2 \times 0.624} = 8.73 \text{ m}$$

$$P_a = 0.5 p_h h = 0.5 \times 61.61 \times 8.73 = 269 \text{ kN}$$

Passive earth pressure:

$$K_p = \tan^2\left(45° + \frac{\phi}{2}\right) = \tan^2\left(45° + \frac{26°}{2}\right) = 2.56$$

$$p_h = K_p \gamma H + 2c\sqrt{K_p} = 2.56 \times 18.1 \times 10 + 2 \times 7.2\sqrt{2.56} = 486.4 \text{ kN/m}$$

$$P_p = 0.5\left[2c\tan\left(45° + \frac{\phi}{2}\right) + p_h\right] H = 0.5(23.04 + 486.4) \times 10 = 2547.2 \text{ kN}$$

$$d_p = \frac{p_h + 4c\tan\left(45° + \frac{\phi}{2}\right)}{3\left[p_h + 2c\tan\left(45° + \frac{\phi}{2}\right)\right]} H = \frac{486.4 + 4 \times 7.2 \times 1.6}{3(486.4 + 2 \times 7.2 \times 1.6)} \times 10 = 3.48 \text{ m}$$

TABLE 10.2 Retaining Structures: Lateral Earth Pressure on Retaining Walls

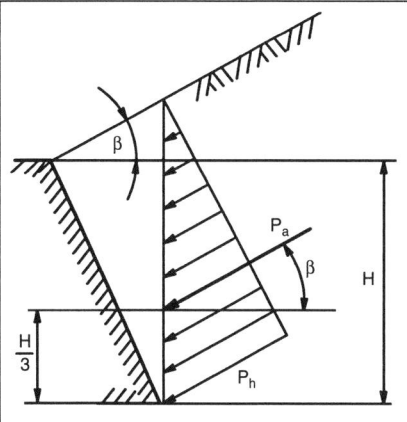

Rankine earth pressure
$P_a = 0.5 K_a \gamma H^2 \quad P_p = 0.5 K_p \gamma H^2$

Rankine theory ($\alpha = 0$, $\delta = 0$)
The wall is assumed to be vertical and smooth.

$$K_a = \cos\beta \frac{\cos\beta - \sqrt{\cos^2\beta - \cos^2\phi}}{\cos\beta + \sqrt{\cos^2\beta - \cos^2\phi}}$$

$$K_p = \cos\beta \frac{\cos\beta + \sqrt{\cos^2\beta - \cos^2\phi}}{\cos\beta - \sqrt{\cos^2\beta - \cos^2\phi}}$$

If $\alpha = 0$, $\delta = 0$, and $\beta = 0$:

$$K_a = \frac{1-\sin\phi}{1+\sin\phi} = \tan^2\left(45° - \frac{\phi}{2}\right)$$

$$K_p = \frac{1+\sin\phi}{1-\sin\phi} = \tan^2\left(45° + \frac{\phi}{2}\right) = \frac{1}{K_a}$$

Examples

1. Assumed: $\alpha = 0$, $\delta = 0$, $\beta = 0$

A: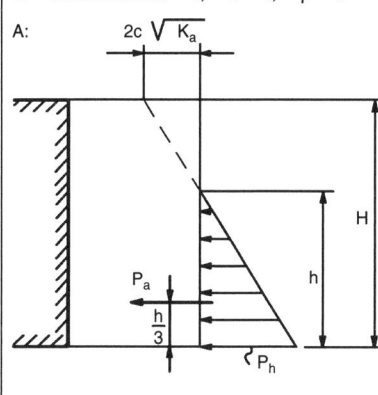

B: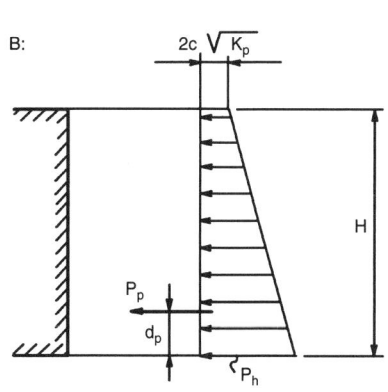

Cohesive soil

A: Active earth pressure

$$p_h = K_a \gamma H - 2c\sqrt{K_a}$$

where c = unit cohesive strength of soil.

$$K_a = \tan^2\left(45° - \frac{\phi}{2}\right), \quad h = \frac{p_h \cdot H}{p_h + 2c\tan(45° - \phi/2)}$$

Resultant force per unit length of wall
$P_a = 0.5 p_h h$

B: Passive earth pressure

$$p_h = K_p \gamma H + 2c\sqrt{K_p}, \quad K_p = \tan^2\left(45° + \frac{\phi}{2}\right)$$

$$P_p = 0.5\left[2c\tan\left(45° + \frac{\phi}{2}\right) + p_h\right] \cdot H$$

$$d_p = \frac{p_h + 4c \cdot \tan\left(45° + \frac{\phi}{2}\right)}{3\left[p_h + 2c \cdot \tan\left(45° + \frac{\phi}{2}\right)\right]} \cdot H$$

─── NOTES ───

Example for Table 10.3. Retaining wall 3 in Table 10.3, $H = 6$ m

Given. Backfill soil: Angle of friction $\phi = 30°$, cohesion $c = 0$
Unit weight of backfill soil $\gamma = 18$ kN/m³
Groundwater: $h_w = 4$ m, $\gamma_w = 9.81$ kN/m³

Required. Compute active pressure per unit length of wall: P_a, d_a.

Solution. $K_a = \tan^2\left(45° - \dfrac{\phi}{2}\right) = \tan^2\left(45° - \dfrac{30°}{2}\right) = 0.333$

$P_1 = 0.5 K_a \gamma (H - h_w)^2 = 0.5 \times 0.333 \times 18 (6-4)^2 = 12.0$ kN

$d_1 = \dfrac{H - h_w}{3} + h_w = \dfrac{6-4}{3} + 4 = 4.67$ m

$P_2 = K_a \gamma (H - h_w) h_w = 0.333 \times 18 (6-4) \times 4 = 48.0$ kN

$d_2 = 0.5 h_w = 0.5 \times 4 = 2$ m

$P_3 = 0.5 K_a (\gamma - \gamma_w) h_w^2 = 0.5 \times 0.333 \times (18 - 9.81) \times 4^2 = 21.8$ kN

$d_3 = \dfrac{h_w}{3} = \dfrac{4}{3} = 1.33$ m

$P_4 = 0.5 \gamma_w h_w^2 = 0.5 \times 9.81 \times 4^2 = 78.5$ kN

$d_4 = \dfrac{h_w}{3} = \dfrac{4}{3} = 1.33$ m

$P_a = P_1 + P_2 + P_3 + P_4 = 12.0 + 48.0 + 21.8 + 78.5 = 160.3$ kN

$d_a = \dfrac{P_1 d_1 + P_2 d_2 + P_3 d_3 + P_4 d_4}{P_a}$

$= \dfrac{12.0 \times 4.67 + 48.0 \times 2 + 21.8 \times 1.33 + 78.5 \times 1.33}{160.3} = 1.78$ m

TABLE 10.3 Retaining Structures: Lateral Earth Pressure on Retaining Walls

2.

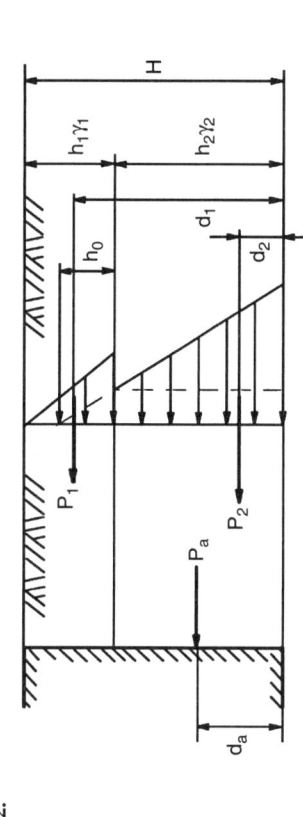

Active earth pressure

$P_1 = 0.5 K_a \gamma_1 h_1^2, \quad d_1 = h_2 + \dfrac{h_1}{3}, \quad h_0 = \dfrac{\gamma_1 h_1}{\gamma_2}$

$P_2 = 0.5 K_a \gamma_2 (2h_0 + h_2) h_2, \quad d_2 = \dfrac{h_2 + 3h_0}{h_2 + 2h_0} \cdot \dfrac{h_2}{3}$

Total active earth pressure $P_a = P_1 + P_2$

$$d_a = \dfrac{P_1 d_1 + P_2 d_2}{P_a}$$

3.

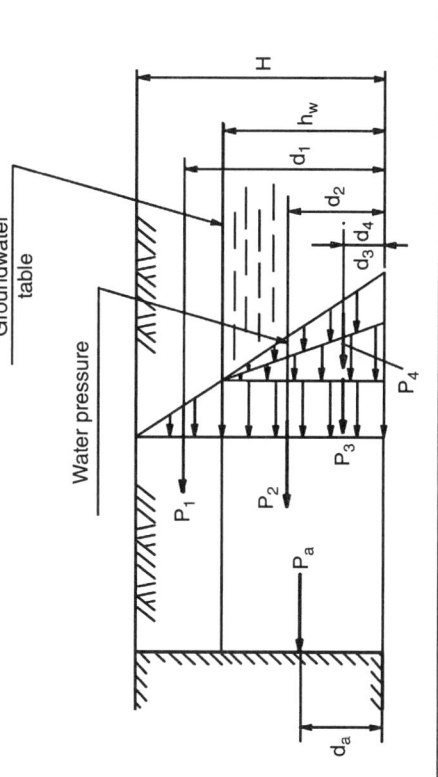

Active earth pressure

$P_1 = 0.5 K_a \gamma (H - h_w)^2, \quad d_1 = \dfrac{H - h_w}{3} + h_w$

$P_2 = K_a \gamma (H - h_w) h_w, \quad d_2 = 0.5 h_w$

$P_3 = 0.5 K_a (\gamma - \gamma_w) h_w^2, \quad d_3 = \dfrac{h_w}{3}$

γ_w = unit weight of water (γ_w = 9.81 kN/m³)

$P_4 = 0.5 \gamma_w h_w^2, \quad d_4 = \dfrac{h_w}{3}$

Total active earth pressure $P_a = P_1 + P_2 + P_3 + P_4$

$$d_a = \dfrac{P_1 d_1 + P_2 d_2 + P_3 d_3 + P_4 d_4}{P_a}$$

NOTES

Cantilever Retaining Walls

Retaining walls having a height ranging from 10 to 20 ft (3.0 to 6.1 m) are generally built as reinforced-concrete cantilever members. As shown in Fig. 10.1, a cantilever wall comprises a vertical stem to retain the soil, a horizontal base to support the stem, and in many instances a key that projects into the underlying soil to augment the resistance to sliding. Adequate drainage is an essential requirement, because the accumulation of water or ice behind the wall would greatly increase the horizontal thrust.

The calculation of earth thrust is based on Rankine's theory.

When a live load, termed a *surcharge*, is applied to the retained soil, it is convenient to replace this load with a hypothetical equivalent prism of earth. Referring to Fig. 10.1, consider a portion QR of the wall, R being at distance y below the top. Take the length of wall normal to the plane of the drawing as 1 ft (0.3 m). Let T = resultant earth thrust on QR; M = moment of this thrust with respect to R; h = height of equivalent earth prism that replaces surcharge; w = unit weight of earth; C_a = coefficient of active earth pressure; C_p = coefficient of passive earth pressure. Then

$$T = \tfrac{1}{2} C_a w y (y + 2h) \qquad (10.1)$$
$$M = \tfrac{1}{6} C_a w y^2 (y + 3h) \qquad (10.2)$$

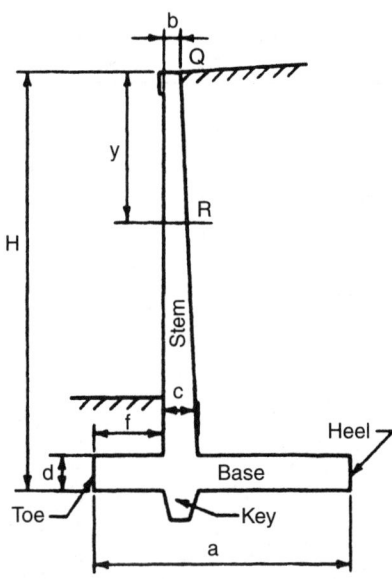

Figure 10.1 Cantilever retaining wall.

TABLE 10.4 Retaining Structures: Lateral Earth Pressure on Retaining Walls

4.	**Active earth pressure** $p_h = K_a\gamma H$, $\quad P_1 = 0.5 K_a \gamma H^2$, $\quad d_1 = \dfrac{H}{3}$ $q_h = K_a w$, $\quad P_2 = 0.5 K_a w H$, $\quad d_2 = \dfrac{H}{2}$ w = uniformly distributed load Total active earth pressure $P_a = P_1 + P_2$ $d_a = \dfrac{P_1 d_1 + P_2 d_2}{P_a} = \dfrac{H + 3w/\gamma}{H + 2w/\gamma} \cdot \dfrac{H}{3}$
5.	**Active earth pressure** $p_h = K_a\gamma H$, $\quad P_1 = 0.5 K_a \gamma H^2$, $\quad d_1 = \dfrac{H}{3}$ $q_h = K_a w$, $\quad P_2 = K_a w h_{q'}$ $h_q = (a+b)\tan\left(45° + \dfrac{\phi}{2}\right) - a\tan\phi$ $d_2 = H - \dfrac{1}{2}\left[(a+b)\tan\left(45° + \dfrac{\phi}{2}\right) + a\tan\phi\right]$ Total active earth pressure $P_a = P_1 + P_2$ $d_a = \dfrac{P_1 d_1 + P_2 d_2}{P_a}$

NOTES

Geosynthetics in Retaining Wall Construction

When the earth adjoining the backfill is a random soil with lower strength than that of the back-fill, the random soil exerts a horizontal pressure on the backfill that is transmitted to the wall (Fig. 10.2). This may lead to a sliding failure of the reinforced zone. The reinforcement at the base should be sufficiently long to prevent this type of failure. The total horizontal sliding force on the base is, from Fig. 10.2,

$$P = P_b + P_s + P_v \tag{10.3}$$

where $P_b = K_a w_b H^2 / 2$
w_b = density of soil adjoining the reinforcement zone
$P_s = K_a w_s h H$
$w_s h$ = weight of uniform surcharge
P_v = force due to live load V as determined by Boussinesq method

The horizontal resisting force is

$$F_H = [(w_s h + w_r H) \tan \phi_{sr} + c] L \tag{10.4}$$

where $w_r H$ = weight of soil in reinforcement zone
ϕ_{sr} = soil reinforcement interaction angle
c = undrained shear strength of backfill
L = length of reinforcement zone base

The safety factor for sliding resistance then is

$$K_{sl} = \frac{F_H}{P} \tag{10.5}$$

and should be 1.5 or larger. A reinforcement length about $0.8H$ generally will provide base resistance sufficient to provide a safety factor of about 1.5.

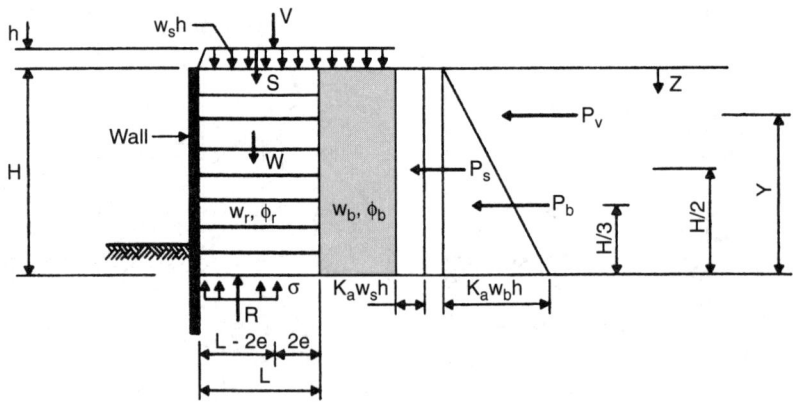

FIGURE 10.2 Retaining wall anchored with geosynthetic reinforcement is subjected to pressure from random soil backfill, sand backfill, surcharge, and live load. Assumed pressure distribution diagrams are rectangular and triangular.

Continued on page 196

TABLE 10.5 Retaining Structures: Lateral Earth Pressure on Retaining Walls

6.

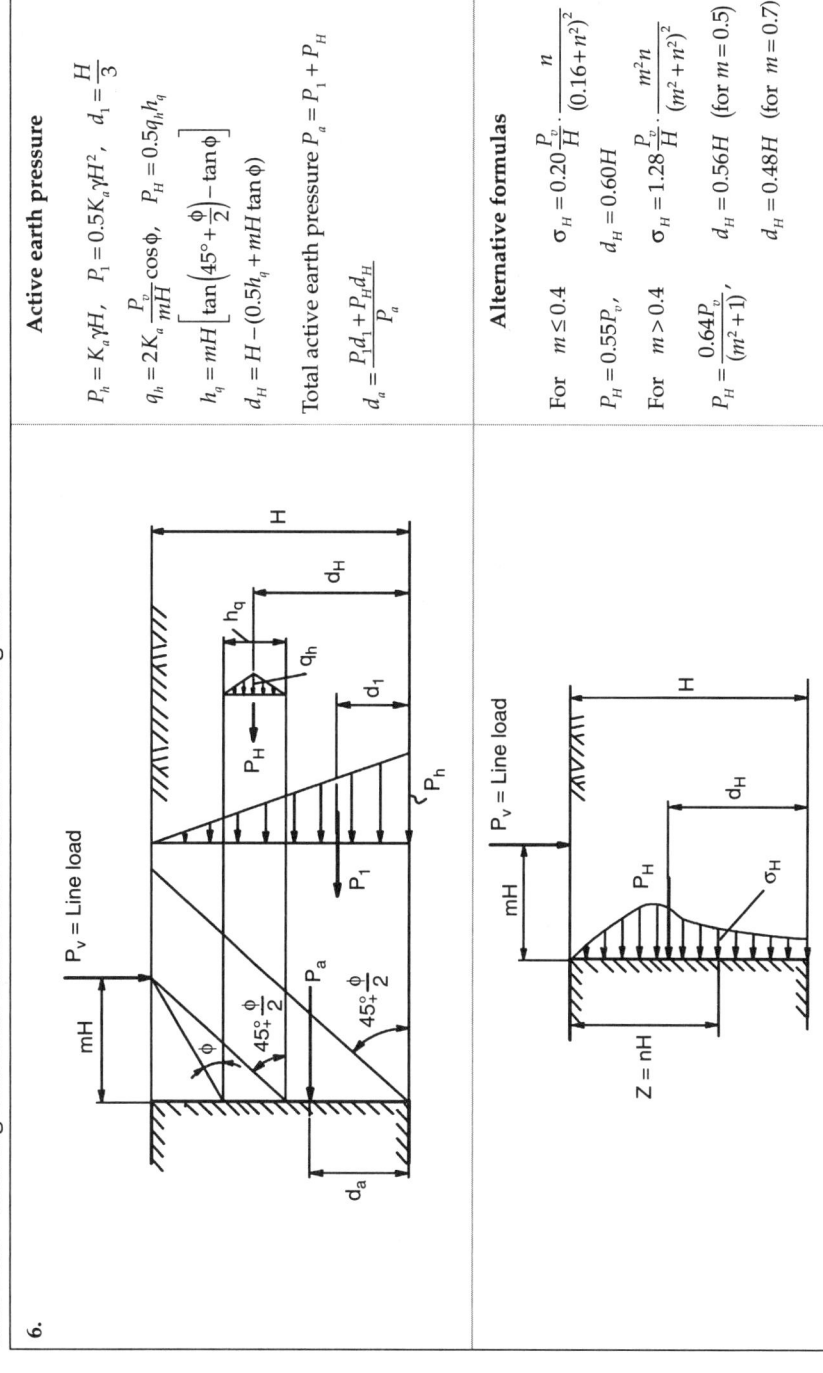

Active earth pressure

$P_h = K_a \gamma H, \quad P_1 = 0.5 K_a \gamma H^2, \quad d_1 = \dfrac{H}{3}$

$q_h = 2K_a \dfrac{P_v}{mH} \cos\phi, \quad P_H = 0.5 q_h h_q$

$h_q = mH \left[\tan\left(45° + \dfrac{\phi}{2}\right) - \tan\phi \right]$

$d_H = H - (0.5 h_q + mH \tan\phi)$

Total active earth pressure $P_a = P_1 + P_H$

$d_a = \dfrac{P_1 d_1 + P_H d_H}{P_a}$

Alternative formulas

For $m \leq 0.4 \quad \sigma_H = 0.20 \dfrac{P_v}{H} \cdot \dfrac{n}{(0.16 + n^2)^2}$

$P_H = 0.55 P_v, \quad d_H = 0.60 H$

For $m > 0.4 \quad \sigma_H = 1.28 \dfrac{P_v}{H} \cdot \dfrac{m^2 n}{(m^2 + n^2)^2}$

$P_H = \dfrac{0.64 P_v}{(m^2 + 1)}, \quad d_H = 0.56 H \quad \text{(for } m = 0.5\text{)}$

$d_H = 0.48 H \quad \text{(for } m = 0.7\text{)}$

> ### NOTES
>
> The most economical retaining wall is one in which the reinforcement is turned upward and backward at the face of the wall and also serves as the face. The backward embedment should be at least 4 ft. If desired for aesthetic reasons or to protect the geosynthetic from damage or deterioration from exposure to ultraviolet light, sprayed concrete may be applied to the wall face.
>
> As an alternative, the wall may be composed of concrete block or precast concrete panels that are anchored to soil reinforcement. The reinforcement should be installed taut to limit lateral movement of the wall during construction.
>
> ## Concrete Gravity Retaining Walls
>
> Forces acting on gravity walls include the weight of the wall, weight of the earth on the sloping back and heel, lateral earth pressure, and resultant soil pressure on the base. It is advisable to include a force at the top of the wall to account for frost action, perhaps 700 lb/lin. ft (1042 kg/m). A wall, consequently, may fail by overturning or sliding, overstressing of the concrete or settlement due to crushing of the soil.
>
> Design usually starts with selection of a trial shape and dimensions, and this configuration is checked for stability. For convenience, when the wall is of constant height, a 1-ft (0.305-m) long section may be analyzed. Moments are taken about the toe. The sum of the righting moments should be at least 1.5 times the sum of the overturning moments. To prevent sliding,
>
> $$\mu R_v \geq 1.5 P_h \qquad (10.6)$$
>
> where μ = coefficient of sliding friction
> R_v = total downward force on soil, lb (N)
> P_h = horizontal component of earth thrust, lb (N)
>
> Next, the location of the vertical resultant R_v should be found at various section of the wall by taking moments about the toe and dividing the sum by R_v. The resultant should act within the middle third of each section if there is to be no tension in the wall.
>
> Finally, the pressure exerted by the base on the soil should be computed to ensure that the allowable pressure is not exceeded. When the resultant is within the middle third, the pressure, lb/ft² (P_a), under the ends of the base are given by
>
> $$p = \frac{R_v}{A} \pm \frac{Mc}{I} = \frac{R_v}{A}\left(1 \pm \frac{6e}{L}\right) \qquad (10.7)$$
>
> where A = area of base, ft² (m²)
> L = width of base, ft (m)
> e = distance, parallel to L, from centroid of base to R_v, ft (m)
>
> Figure 10.3b shows the pressure distribution under a 1-ft (0.305-m) strip of wall for $e = L/2 - a$, where a is the distance of R_v from the toe. When R_v is exactly $L/3$ from the toe, the pressure at the heel becomes zero. When R_v falls outside the middle third, the pressure vanishes under a zone around the heel, and pressure at the toe is much larger than for the other cases.

Continued on page 198

TABLE 10.6 Retaining Structures: Lateral Earth Pressure on Braced Sheetings

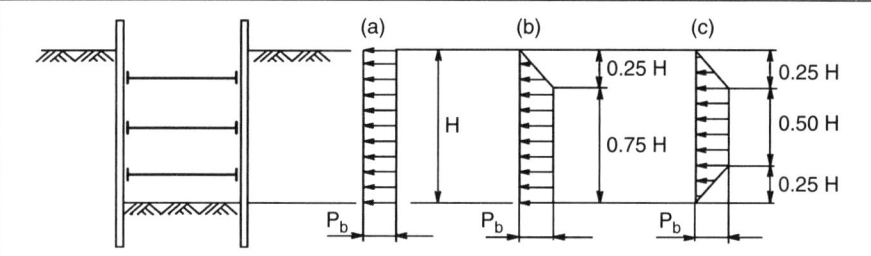

Empirical diagrams of lateral earth pressure on braced sheetings

a: Sand: $P_b = 0.65 \gamma H \tan^2\left(45° - \dfrac{\phi}{2}\right)$

b: Soft to medium clay: $P_b = \gamma H - 2q_u$, q_u = unconfined compressive strength, $q_u = 2c$

c: Stiff-fissured clay: $P_b = 0.2\gamma H$ to $0.4\gamma H$

Lateral Earth Pressure on Basement Walls

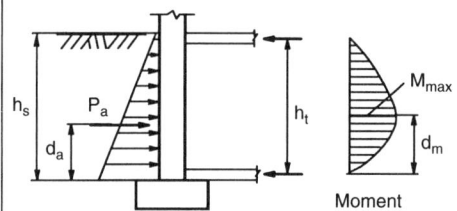

Active earth pressure:
$$P_a = 0.5 K \gamma h_s^2, \quad d_a = \dfrac{h_s}{3}$$

Maximum bending moment:
$$M_{max} = 0.128 P_a h_s, \quad d_m = 0.42 h_s$$

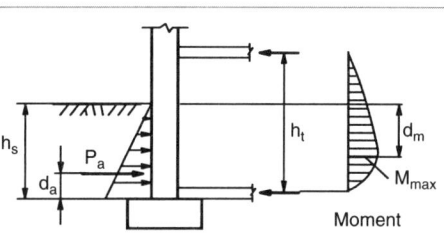

Active earth pressure:
$$P_a = 0.5 K_a \gamma h_s^2, \quad d_a = \dfrac{h_s}{3}$$

Maximum bending moment:
$$M_{max} = \dfrac{P_a h_s}{3 h_t}\left(h + \dfrac{2 h_s}{3}\sqrt{\dfrac{h_s}{3 h_t}}\right), \quad d_m = h_s \sqrt{\dfrac{h_s}{3 h_t}}$$

NOTES

The variables in the five formulas in Fig. 10.3 are

P_1 and P_2 = pressure, lb/ft² (MPa), at locations shown
L and a = dimensions, ft (m), at locations shown
R_v = total downward force on soil behind retaining wall, lb (N)
R = resultant, lb (N)

In usual design work on retaining walls, the sum of the righting moments and the sum of the overturning moments about the toe are found. It is assumed by designers that if the retaining wall is overturned, it will overturn about the toe of the retaining wall. Designers then apply a safety factor thus:

Retaining wall righting moment = 1.5 (overturning moment)

The 1.5 safety factor is a common value among designers.

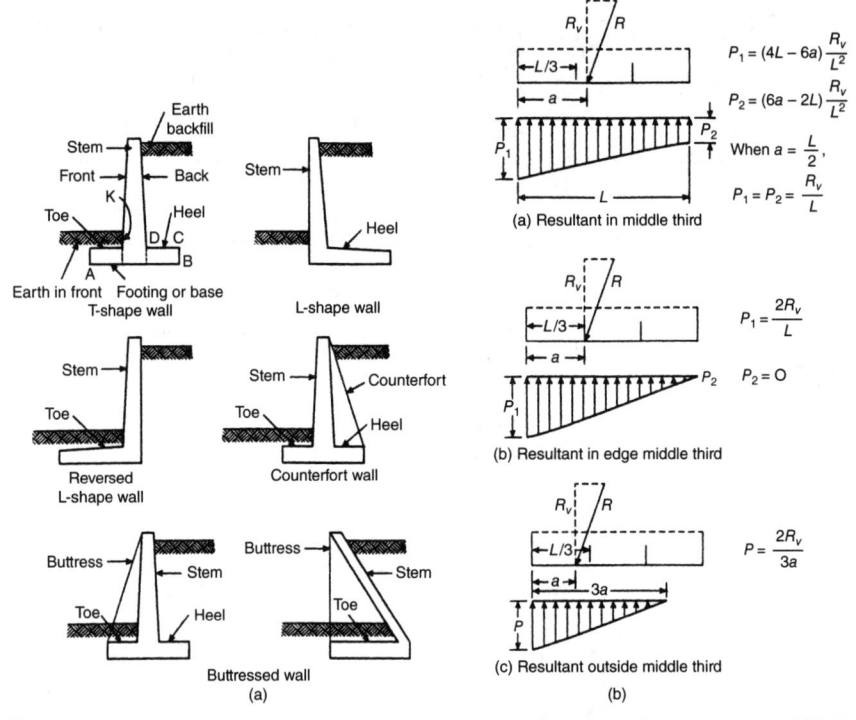

FIGURE 10.3 (a) Six types of retaining walls. (b) Soil-pressure variations in retaining walls. (Merritt-*Building Construction Handbook*, McGraw-Hill.)

TABLE 10.7 Retaining Structures: Cantilever Retaining Walls

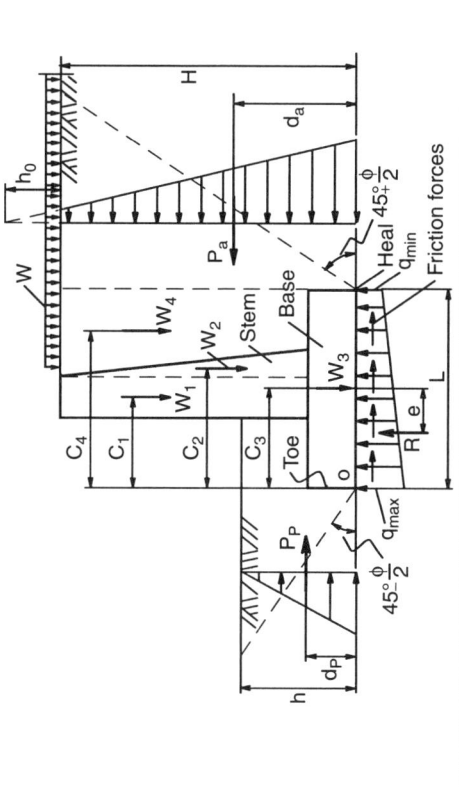

Stability analysis

W_i = weight (concentrated load for width $B = 1$)
w = surcharge (uniformly distributed load), $h_0 = w/\gamma$

Active earth pressure:

$$P_a = 0.5\gamma H \tan^2\left(45° - \frac{\phi}{2}\right)(H + 2h_0), \quad d_a = \frac{H}{3} \cdot \frac{H + 3h_0}{H + 2h_0}$$

Passive earth pressure:

$$P_p = 0.5\gamma H \tan^2\left(45° + \frac{\phi}{2}\right), \quad d_p = \frac{h}{3}$$

The factor of safety against sliding

$$F.S. = \frac{\text{resisting force } F}{\text{actual horizontal force } \sum P_H}$$

where $F = f\sum W_i$, $\sum P_H = P_a - P_p$

f = coefficient of friction ($f = 0.4$ to 0.5)

F.S. = 1.5 to 2.0

The factor of safety against overturning

$$F.S. = \frac{\text{stabilizing moment about toe } \left(\sum M_r\right)}{\text{overturning moment about toe } \left(\sum M_o\right)}$$

where $\sum M_r = \sum W_i c_i + P_p d_p$, $\sum M_o = P_a d_a$

F.S. = 1.5 to 2.0

The factor of safety of bearing capacity failure

$$F.S. = \frac{\text{soil's ultimate bearing capacity}}{\text{maximum contact (base) pressure}}, \quad F.S. = 3.0$$

Eccentricity of resultant force R: $\quad e = \frac{L}{2} - \frac{\sum M_o}{\sum W_i} \leq \frac{L}{6}, \quad R = \sum W_i$

Maximum contact (base) pressure: $\quad q_{max} = \frac{\sum W_i}{L \cdot B} + \frac{6\sum W_i \cdot e}{L^2 \cdot B}, \quad (B = 1)$

─── NOTES ───

Example for Table 10.8. Cantilever sheet piling 2 in Table 10.8, $H = 10$ m

Given. Soil properties: $\phi_1 = 32°$, $c_1 = 0$, $\gamma_1 = 18$ kN/m³
$\phi_2 = 34°$, $c_2 = 0$, $\gamma_2 = 16$ kN/m³, $\beta = 0$, $\alpha = 0$, $\delta = 0$

Required. Compute depth D and maximum bending moment M_{max} per unit length of sheet piling.

Solution. $K_{a_1} = \tan^2\left(45° - \dfrac{\phi_1}{2}\right) = \tan^2\left(45° - \dfrac{32°}{2}\right) = 0.307$

$K_{a_2} = \tan^2\left(45° - \dfrac{\phi_2}{2}\right) = \tan^2\left(45° - \dfrac{34°}{2}\right) = 0.283$

$K_{p_2} = \tan^2\left(45° + \dfrac{\phi_2}{2}\right) = \tan^2\left(45° + \dfrac{34°}{2}\right) = 3.537$, $K_{p_2} - K_{a_2} = 3.254$

$P_1 = 0.5 K_{a_1} \gamma_1 H^2 = 0.5 \times 0.307 \times 18 \times 10^2 = 276.3$ kN,

$z_1 = \dfrac{K_{a_2} \gamma_1 H}{(K_{p_2} - K_{a_2})\gamma_2} = \dfrac{0.283 \times 18 \times 10}{3.254 \times 16} = 0.98$ m

$P_2 = 0.5 K_{a_2} \gamma_1 H z_1 = 0.5 \times 0.283 \times 18 \times 10 \times 0.98 = 24.96$ kN,

$z_2 = \sqrt{\dfrac{P_1 + P_2}{0.5(K_{p_2} - K_{a_2})\gamma_2}} = \sqrt{\dfrac{276.3 + 24.96}{0.5 \times 3.254 \times 16}} = 3.4$ m

$P_3 = 0.5(K_{p_2} - K_{a_2})\gamma_2 (D_0 - z_1)^2 = 0.5 \times 3.254 \times 16 (D_0 - z_1)^2 = 26.03(D_0 - z_1)^2$

$\sum M_d = 0$ (condition of equilibrium)

$P_1\left(\dfrac{H}{3} + D_0\right) + P_2\left(D_0 - \dfrac{z_1}{3}\right) - P_3 \dfrac{1}{3}(D_0 - z_1) = 0$

$276.3\left(\dfrac{10}{3} + D_0\right) + 24.96 D_0 - 26.03 \dfrac{1}{3}(D_0 - z_1)^3 = 0$

$8.68(D_0 - z_1)^3 = 921.0 + 301.26 D_0$

Using method of trial and error:

Assume $D_0 = 8.3$ m, $(8.3 - 0.98)^3 = 106.10 + 34.71 \times 8.3$, $394.19 \approx 393.18$

$D = 1.2 D_0 = 1.2 \times 8.3 = 9.96$ m

$M_{max} = P_1\left(\dfrac{H}{3} + z_1 + z_2\right) + P_2\left(\dfrac{2}{3} z_1 + z_2\right) - 0.5(K_{p_2} - K_{a_2})\gamma_2 z_2^2 \left(\dfrac{z_2}{3}\right)$

$= 276.3\left(\dfrac{10}{3} + 0.98 + 3.4\right) + 24.96\left(\dfrac{2}{3} \times 0.98 + 3.4\right)$

$- 0.5 \times 3.254 \times 16 \times 3.4^2 \left(\dfrac{3.4}{3}\right) = 1891.4$ kN·m/m

TABLE 10.8 Retaining Structures: Cantilever Sheet Pilings

1. 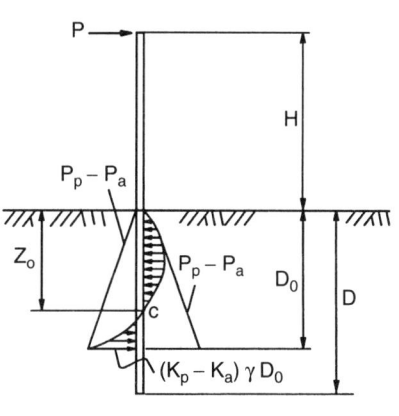	Equation to determine the embedment (D_0): $$P = \frac{(K_p - K_a)\gamma D_0^3}{6(4H + 3D_0)}$$ Maximum bending moment: $$M_{max} = P\left[H + \frac{2}{3}\sqrt{\frac{P}{(K_p - K_a)\gamma}}\right]$$ $$z_c = D_0 \frac{4H + 3D_0}{6H + 4D_0}$$ For single pile $$P = \frac{(K_p - K_a)\gamma d D_0^3}{3(4H + 3D_0)},$$ $$M_{max} = P\left(H + \frac{2}{3}\sqrt{\frac{P}{(K_p - K_a)\gamma d}}\right),$$ where d = pile diameter $D = (1.2 \text{ to } 1.4) D_0$ for factor of safety at 1.5 to 2.0
2. 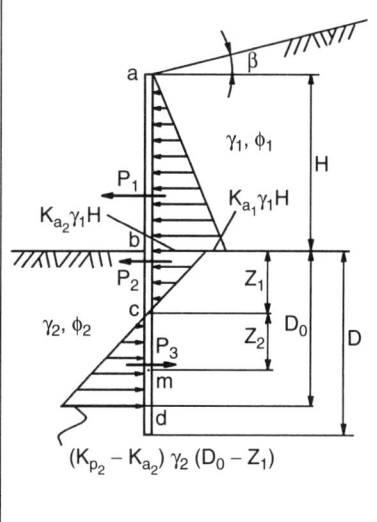	Earth pressure: $$P_1 = 0.5 K_{a_1}\gamma H^2, \quad z_1 = \frac{K_{a_1}\gamma_1 H}{(K_{p_2} - K_{a_2})\gamma_2}$$ $$P_2 = 0.5 K_{a_2}\gamma_1 H \cdot z_1, \quad z_2 = \sqrt{\frac{P_1 + P_2}{0.5(K_{p_2} - K_{a_2})\gamma_2}}$$ $$P_3 = 0.5(K_{p_2} - K_{a_2})\gamma_2 (D_0 - z_1)^2$$ Equation to determine D_0: $\sum M_d = 0$ $$P_1\left(\frac{H}{3} + D_0\right) + P_2\left(D_0 - \frac{z_1}{3}\right) - P_3 \frac{1}{3}(D_0 - z_1) = 0$$ $D = (1.2 \text{ to } 1.4) D_0$ for factor of safety at 1.5 to 2.0 m = point of zero shear and maximum bending moment Maximum bending moment $$M_{max} = P_1\left(\frac{H}{3} + z_1 + z_2\right) + P_2\left(\frac{2}{3}z_1 + z_2\right)$$ $$- 0.5(K_p - K_a)\gamma z_2^2 \cdot \left(\frac{z_2}{3}\right)$$

─── NOTES ───

Example for Table 10.9. Anchored sheet pile wall in Table 10.9, $H = 15$ m

Given. Soil properties: $\phi_1 = 30°$, $c_1 = 0$, $\gamma_1 = 20$ kN/m³,
$\phi_2 = 32°$, $c_2 = 0$, $\gamma_2 = 18$ kN/m³,
$\beta = 0$, $\alpha = 0$, $\delta = 0$, $d = 1.2$ m

Required. Compute depth D and maximum bending moment M_{max} per unit length of wall.

Solution.
$$K_{a_1} = \tan^2\left(45° - \frac{\phi_1}{2}\right) = \tan^2\left(45° - \frac{30°}{2}\right) = 0.333,$$

$$K_{a_2} = \tan^2\left(45° - \frac{\phi_2}{2}\right) = \tan^2\left(45° - \frac{32°}{2}\right) = 0.307,$$

$$K_{p_2} = \tan^2\left(45° + \frac{\phi_2}{2}\right) = \tan^2\left(45° + \frac{32°}{2}\right) = 3.254, \quad K_{p_2} - K_{a_2} = 2.948$$

Forces per unit length of wall

$P_1 = 0.5 K_{a_1} \gamma_1 d^2 = 0.5 \times 0.333 \times 20 \times 1.2^2 = 4.8$ kN

$P_2 = 0.5 K_{a_1} \gamma_1 (H+d)(H-d) = 0.5 \times 0.333 \times 20 \times (15+1.2)(15-1.2) = 744.4$ kN

$$d_2 = \frac{(H-d)(2H+d)}{3(H+d)} = \frac{(15-1.2)(2 \times 15 + 1.2)}{3(15+1.2)} = 8.86 \text{ m}$$

$P_3 = 0.5 K_{a_2} \gamma_1 H z_1 = 0.5 \times 0.307 \times 20 \times 15 \times 1.74 = 80.13$ kN

$$z_1 = \frac{K_{a_2} \gamma_1 H}{(K_{p_2} - K_{a_2})\gamma_2} = \frac{0.307 \times 20 \times 15}{2.948 \times 18} = 1.74 \text{ m}$$

For $\phi_2 = 32°$: $x = 0.059 H = 0.059 \times 15 = 0.885$

$$\sum M_T = 0, \quad R(H - d + x) + P_1 \frac{d}{3} - P_2 d_2 - P_3 \left(H - d + \frac{z_1}{3}\right) = 0$$

$$R(15 - 1.2 + 0.885) + 4.8 \times \frac{1.2}{3} - 744.4 \times 8.86 - 80.13\left(15 - 1.2 + \frac{1.74}{3}\right) = 0,$$

$R = 527.46$ kN

$T = (P_1 + P_2 + P_3) - R = 4.8 + 744.4 + 80.13 - 527.46 = 301.87$ kN

$$D_0 = z_1 + \sqrt{\frac{6R}{(K_{p_2} - K_{a_2})\gamma_2}} = 1.74 + \sqrt{\frac{6 \times 301.87}{2.948 \times 18}} = 7.58 \text{ m}, \quad \text{(assumed } x = z_1\text{)}$$

$D = 1.2 D_0 = 1.2 \times 7.58 = 9.1$ m

$$z_2 = \sqrt{\frac{P_1 + P_2 + P_3 - T}{0.5(K_{p_2} - K_{a_2})\gamma_2}} = \sqrt{\frac{4.8 + 744.4 + 80.13 - 301.87}{0.5 \times 2.948 \times 18}} = 4.46 \text{ m}$$

$$M_{max} = (P_1 + P_2)\left(\frac{H}{3} + z_1 + z_2\right) + P_3\left(\frac{2}{3} z_1 + z_2\right) - T(H - d + z_1 + z_2) - 0.5(K_{p_2} - K_{a_2})\gamma_2 z_2^2 \left(\frac{z_2}{3}\right)$$

$$= (4.8 + 744.4)\left(\frac{15}{3} + 1.74 + 4.46\right) + 80.13\left(\frac{2}{3} \times 1.74 + 4.46\right) - 301.87(15 - 1.2 + 1.74 + 4.46)$$

$$- 0.5 \times 2.948 \times 18 \times \frac{4.46^3}{3} = 2019.4 \text{ kN} \cdot \text{m/m}$$

TABLE 10.9 Retaining Structures: Anchored Sheet Pile Walls

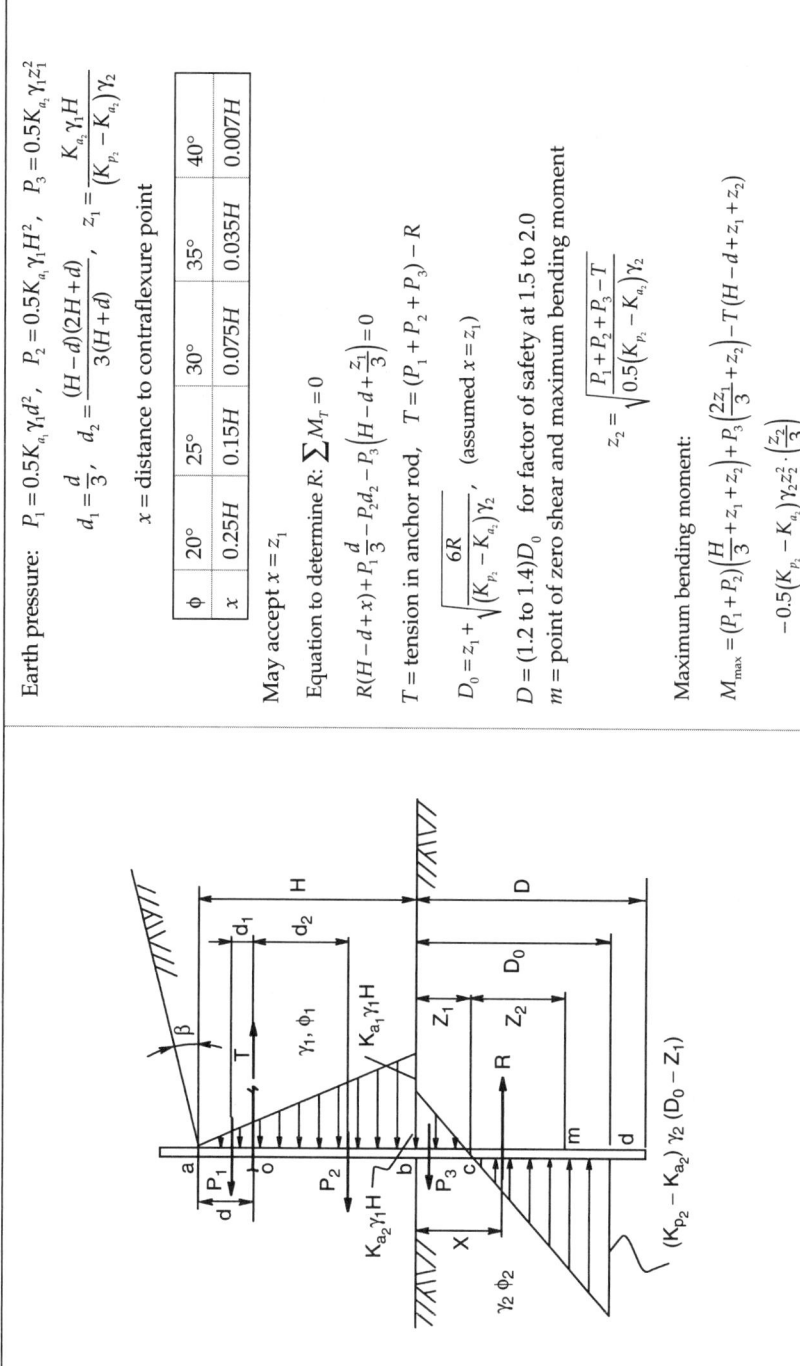

Earth pressure: $P_1 = 0.5 K_{a_1} \gamma_1 d^2$, $P_2 = 0.5 K_{a_1} \gamma_1 H^2$, $P_3 = 0.5 K_{a_2} \gamma_1 z_1^2$

$$d_1 = \frac{d}{3}, \quad d_2 = \frac{(H-d)(2H+d)}{3(H+d)}, \quad z_1 = \frac{K_{a_2}\gamma_1 H}{(K_{p_2} - K_{a_2})\gamma_2}$$

x = distance to contraflexure point

ϕ	20°	25°	30°	35°	40°
x	0.25H	0.15H	0.075H	0.035H	0.007H

May accept $x = z_1$

Equation to determine R: $\sum M_T = 0$

$$R(H - d + x) + P_1 \frac{d}{3} - P_2 d_2 - P_3\left(H - d + \frac{z_1}{3}\right) = 0$$

T = tension in anchor rod, $\quad T = (P_1 + P_2 + P_3) - R$

$$D_0 = z_1 + \sqrt{\frac{6R}{(K_{p_2} - K_{a_2})\gamma_2}}, \quad \text{(assumed } x = z_1\text{)}$$

$D = (1.2 \text{ to } 1.4) D_0 \quad$ for factor of safety at 1.5 to 2.0

m = point of zero shear and maximum bending moment

$$z_2 = \sqrt{\frac{P_1 + P_2 + P_3 - T}{0.5(K_{p_2} - K_{a_2})\gamma_2}}$$

Maximum bending moment:

$$M_{\max} = (P_1 + P_2)\left(\frac{H}{3} + z_1 + z_2\right) + P_3\left(\frac{2z_1}{3} + z_2\right) - T(H - d + z_1 + z_2)$$
$$- 0.5(K_{p_2} - K_{a_2})\gamma_2 z_2^2 \cdot \left(\frac{z_2}{3}\right)$$

Stability of a Retaining Wall

Determine the factor of safety (F.S.) against sliding and overturning of the concrete retaining wall in Fig. 10.4. The concrete weighs 150 lb/ft³ (23.56 kN/m³), the earth weighs 100 lb/ft³ (15.71 kN/m³), the coefficient of friction is 0.6, and the coefficient of active earth pressure is 0.333.

Calculation Procedure

1. Compute the vertical loads on the wall.
Select a 1-ft (304.8-mm) length of wall as typical of the entire structure. The horizontal pressure of the confined soil varies linearly with the depth and is represented by the triangle BGF in Fig. 10.4.

Resolve the wall into the elements $AECD$ and AEB; pass the vertical plane BF through the soil. Calculate the vertical loads, and locate their resultants with respect to the toe C. Thus $W_1 = 15(1)(150) = 2250$ lb (10,008 N); $W_2 = 0.5(15)(5) \times (150) = 5625$; $W_3 = 0.5(15)(5)(100) = 3750$. Then $\Sigma W = 11{,}625$ lb (51,708 N). Also, $x_1 = 0.5$ ft; $x_2 = 1 + 0333(5) = 2.67$ ft (0.81 m); $x_3 = 1 + 0.667(5) = 433$ ft (1.32 m).

2. Compute the resultant horizontal soil thrust.
Compute the resultant horizontal thrust T lb of the soil by applying the coefficient of active earth pressure. Determine the location of T. Thus $BG = 0.333(15)(100) = 500$ lb/lin. ft (7295 N/m); $T = 0.5(15)(500) = 3750$ lb (16,680 N); $y = 0.333(15) = 5$ ft (1.5 m).

3. Compute the maximum frictional force preventing sliding.
The maximum frictional force $F_m = \mu(\Sigma W)$, where μ = coefficient of friction. Or $F_m = 0.6(11{,}625) = 6975$ lb (31,024.8 N).

4. Determine the factor of safety against sliding.
The factor of safety against sliding is F.S.S. $= F_m/T = 6975/3750 = 1.86$.

5. Compute the moment of the overturning and stabilizing forces.
Taking moments with respect to C, we find the overturning moment $= 3750\,(5) = 18{,}750$ lb · ft (25,406.3 N · m). Likewise, the stabilizing moment $= 2250(0.5) + 5625\,(2.67) + 3750(4.33) = 32{,}375$ lb · ft (43,868.1 N · m).

6. Compute the factor of safety against overturning.
The factor of safety against overturning is F.S.O. = stabilizing moment, lb · ft (N · m)/overturning moment, lb · ft (N · m) $= 32{,}375/18{,}750 = 1.73$.

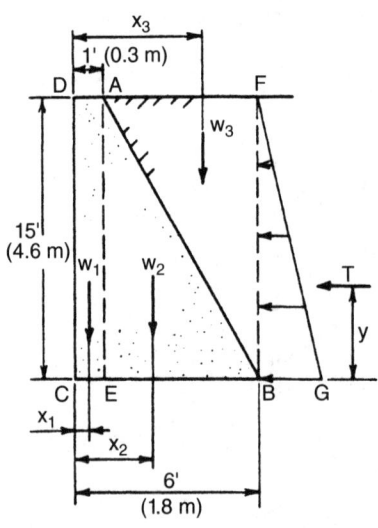

FIGURE 10.4 Concrete retaining wall.

CHAPTER 11

Pipes and Tunnels: Bending Moments for Various Static Loading Conditions

206 Retaining Structures, Pipes, and Tunnels

NOTES

This chapter provides formulas for computation of bending moments in various structures with rectangular or circular cross sections, including underground pipes and tunnels. The formulas for structures with circular cross sections can also be used to compute axial forces and shears.

The formulas provided are applicable to the analysis of elastic systems only. The tables contain the most common cases of loading conditions.

TABLE 11.1 Pipes and Tunnels: Rectangular Cross Section

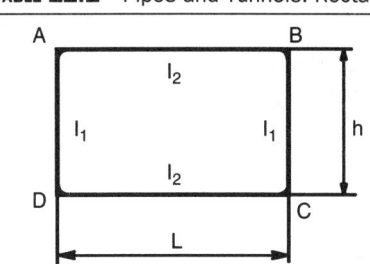

$$k = \frac{I_2 h}{I_1 L}$$

$+M$ = tension on inside of section

1.

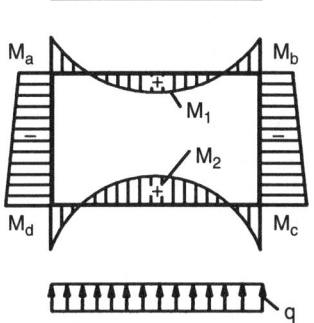

For $q \neq w$

$$M_a = M_b = -\frac{L^2}{12} \cdot \frac{w(2k+3) - qk}{k^2 + 4k + 3}$$

$$M_c = M_d = -\frac{L^2}{12} \cdot \frac{q(2k+3) - wk}{k^2 + 4k + 3}$$

For $q = w$

$$M_a = M_b = M_c = M_d = -\frac{wL^2}{12} \cdot \frac{k+3}{k^2 + 4k + 3}$$

2.

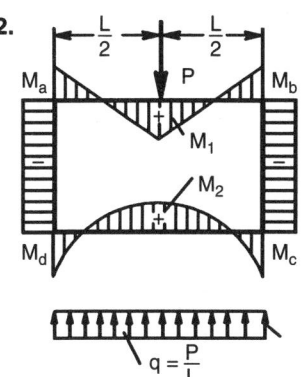

$$M_a = M_b = -\frac{PL}{24} \cdot \frac{4k+9}{k^2 + 4k + 3}$$

$$M_c = M_d = -\frac{PL}{24} \cdot \frac{4k+6}{k^2 + 4k + 3}$$

For $k = 1$

$$M_a = M_b = -\frac{13}{192} PL$$

$$M_c = M_d = -\frac{7}{192} PL$$

─── NOTES ───

Pressure on Submerged Curved Surfaces

The hydrostatic pressure on a submerged curved surface (Fig. 11.1) is given by

$$P = \sqrt{P_H^2 + P_V^2} \tag{11.1}$$

where P = total pressure force on surface
  P_H = force due to pressure horizontally
  P_V = force due to pressure vertically

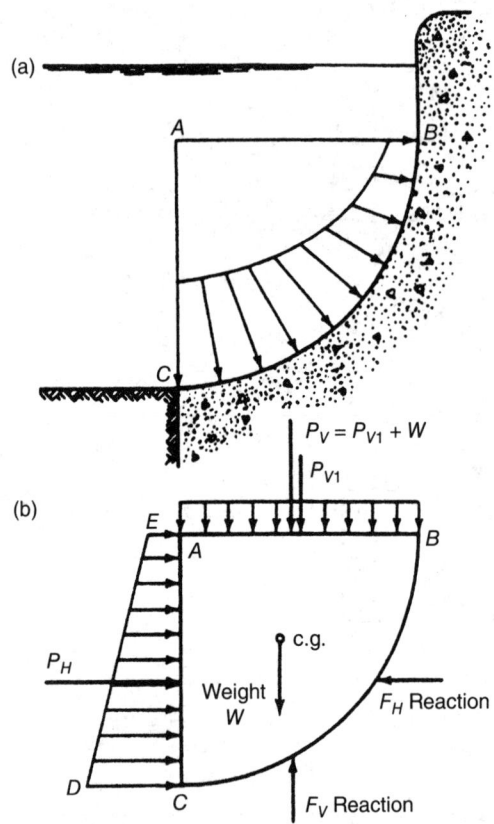

FIGURE 11.1 Hydrostatic pressure on a submerged curved surface: (a) Pressure variation over the surface. (b) Free-body diagram.

Pipes and Tunnels

TABLE 11.2 Pipes and Tunnels: Rectangular Cross Section

3. 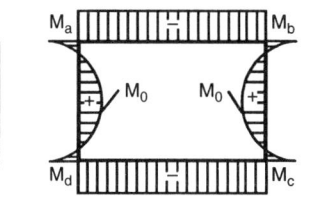	$M_a = -\dfrac{ph^2 k}{12(k+1)}$ $M_b = M_c = M_d = M_a$ For $k = 1$ and $h = L$ $M_a = M_b = M_c = M_d = -\dfrac{ph^2}{24}$ $M_0 = 0.125 ph^2 - 0.5(M_a + M_d)$
4. 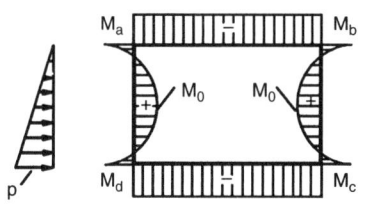	$M_a = M_b = -\dfrac{ph^2 k(2k+7)}{60(k^2 + 4k + 3)}$ $M_c = M_d = -\dfrac{ph^2 k(3k+8)}{60(k^2 + 4k + 3)}$ For $k = 1$ and $h = L$ $M_a = M_b = -\dfrac{3ph^2}{160}, \quad M_c = M_d = -\dfrac{11 ph^2}{480}$ $M_0 = 0.064 ph^2 - [M_a + 0.577(M_d - M_a)]$
5. 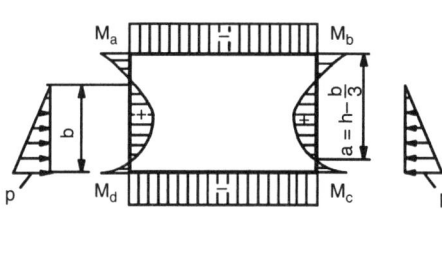	$M_a = M_b = -\dfrac{(A+D)(2k+3) - D(3k+3)}{3(k^2 + 4k + 3)}$ $M_c = M_d = -\dfrac{D(3k+3) - (A+D)k}{3(k^2 + 4k + 3)}$ $A = \dfrac{pb^2 k}{60 h^2}(10 h^2 - 3 b^2)$ $D = \dfrac{pbak}{2h^2}\left(h^2 - a^2 - b^2 \dfrac{45a - 2b}{270 a}\right)$

NOTES

Example for Table 11.3 Rectangular pipe 7 in Table 11.3

Given. Concrete frame, $L = 4$ m, $H = 2.5$ m, $h_1 = 10$ cm, $h_2 = 20$ cm
$b = 1$ m (unit length of pipe)

$$I_1 = \frac{bh_1^3}{12} = \frac{100 \times 10^3}{12} = 8333 \text{ cm}^4, \quad I_2 = \frac{bh_2^3}{12} = \frac{100 \times 20^3}{12} = 66{,}667 \text{ cm}^4$$

Uniformly distrubuted load $w = 120$ kN/m

Required. Compute bending moments.

Solution. $k = \dfrac{I_2 H}{I_1 L} = \dfrac{66{,}667 \times 2.5}{8333 \times 4} = 5.0, \quad r = 2k + 1 = 2 \times 5 + 1 = 11$

$m = 20(k+2)m = 20(k+2)(6k^2 + 6k + 1) = 20(5+2)(6 \times 5^2 + 6 \times 5 + 1) = 25{,}340$

$\alpha_1 = 138k^2 + 265k + 43 = 138 \times 5^2 + 265 \times 5 + 43 = 4818$

$\alpha_2 = 78k^2 + 205k + 33 = 78 \times 5^2 + 205 \times 5 + 33 = 3008$

$\alpha_3 = 81k^2 + 148k + 37 = 81 \times 5^2 + 148 \times 5 + 37 = 2802$

$\alpha_4 = 21k^2 + 88k + 27 = 21 \times 5^2 + 88 \times 5 + 27 = 992$

$$M_a = \frac{wL^2}{24}\left(\frac{1}{r} + \frac{\alpha_1}{m}\right) = \frac{120 \times 4^2}{24}\left(\frac{1}{11} + \frac{4818}{25{,}340}\right) = -22.56 \text{ kN} \cdot \text{m},$$

$$M_e = -\frac{wL^2}{24}\left(\frac{1}{r} - \frac{\alpha_1}{m}\right) = +7.92 \text{ kN} \cdot \text{m}$$

$$M_c = -\frac{wL^2}{24}\left(\frac{1}{r} + \frac{\alpha_2}{m}\right) = \frac{120 \times 4^2}{24}\left(\frac{1}{11} + \frac{3008}{25{,}340}\right) = -16.78 \text{ kN} \cdot \text{m},$$

$$M_f = -\frac{wL^2}{24}\left(\frac{1}{r} - \frac{\alpha_2}{m}\right) = +2.24 \text{ kN} \cdot \text{m}$$

$$M_{b1} = -\frac{wL^2}{24}\left(\frac{3k+1}{r} + \frac{\alpha_2}{m}\right) = \frac{120 \times 4^2}{24}\left(\frac{3 \times 5}{11} + \frac{2802}{25{,}340}\right) = -125.2 \text{ kN} \cdot \text{m}$$

$$M_{b2} = -\frac{wL^2}{24}\left(\frac{3k+1}{r} - \frac{\alpha_3}{m}\right) = -107.44 \text{ kN} \cdot \text{m},$$

$$M_{b4} = -\frac{wL^2}{12} \cdot \frac{\alpha_2}{m} = -17.76 \text{ kN} \cdot \text{m}$$

$$M_{d4} = -\frac{wL^2}{12} \cdot \frac{\alpha_4}{m} = -\frac{120 \times 4^2}{12} \cdot \frac{992}{25{,}340} = -6.24 \text{ kN} \cdot \text{m}$$

$$M_{d6} = -\frac{wL^2}{24}\left(\frac{3k+1}{r} - \frac{\alpha_4}{m}\right) = -\frac{120 \times 4^2}{24}\left(\frac{3 \times 5 + 1}{11} + \frac{992}{25{,}340}\right) = -119.44 \text{ kN} \cdot \text{m}$$

TABLE 11.3 Pipes and Tunnels: Rectangular Cross Section

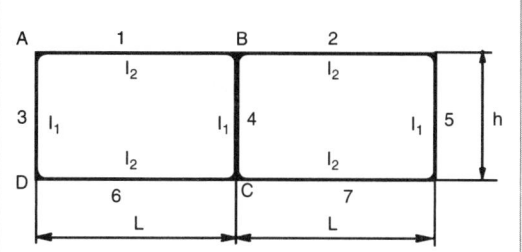

$$k = \frac{I_2 h}{I_1 L}$$

$$r = 2k + 1$$

$$m = 20(k+2)(6k^2 + 6k + 1)$$

$+M$ = tension on inside of section

6.

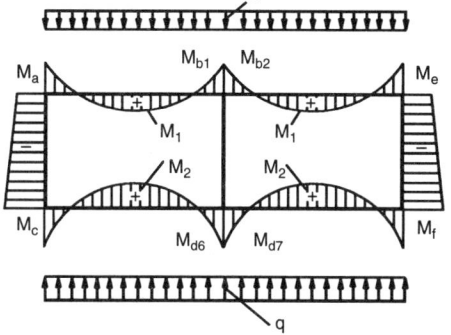

$q = w$

$$M_a = -\frac{wL^2}{12} \cdot \frac{1}{r}, \quad M_c = M_e = M_f = M_a$$

$$M_{b1} = -\frac{wL^2}{12} \cdot \frac{3k+1}{r}, \quad M_{b2} = M_{d6} = M_{d7} = M_{b1}$$

$$M_{b4} = M_{d4} = 0$$

7.

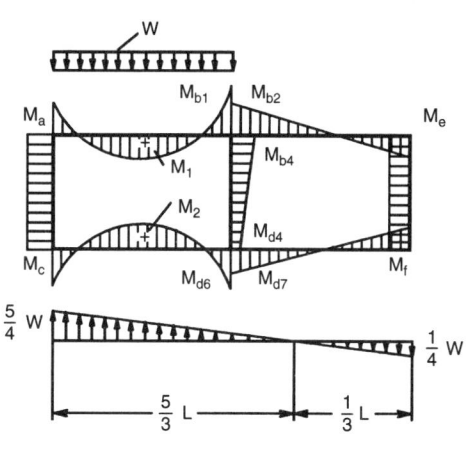

$$M_a = -\frac{wL^2}{24}\left(\frac{1}{r} + \frac{\alpha_1}{m}\right), \quad M_e = -\frac{wL^2}{24}\left(\frac{1}{r} - \frac{\alpha_1}{m}\right)$$

$$M_c = -\frac{wL^2}{24}\left(\frac{1}{r} + \frac{\alpha_2}{m}\right), \quad M_f = -\frac{wL^2}{24}\left(\frac{1}{r} - \frac{\alpha_2}{m}\right)$$

$$M_{b1} = -\frac{wL^2}{24}\left(\frac{3k+1}{r} + \frac{\alpha_3}{m}\right), \quad M_{b4} = -\frac{wL^2}{12} \cdot \frac{\alpha_3}{m}$$

$$M_{b2} = -\frac{wL^2}{24}\left(\frac{3k+1}{r} - \frac{\alpha_3}{m}\right), \quad M_{d4} = -\frac{wL^2}{12} \cdot \frac{\alpha_4}{m}$$

$$M_{d6} = -\frac{wL^2}{24}\left(\frac{3k+1}{r} + \frac{\alpha_4}{m}\right)$$

$$M_{d7} = -\frac{wL^2}{24}\left(\frac{3k+1}{r} - \frac{\alpha_4}{m}\right)$$

$\alpha_1 = 138k^2 + 265k + 43, \quad \alpha_3 = 81k^2 + 148k + 37$

$\alpha_2 = 78k^2 + 205k + 33, \quad \alpha_4 = 21k^2 + 88k + 27$

--- NOTES ---

Loads on Ditch Piping and Conduit

The load, lb/lin. ft, on a rigid ditch conduit may be computed from

$$W = C_D whb \qquad (11.2)$$

and on a flexible ditch conduit from

$$W = C_D whD \qquad (11.3)$$

where C_D = load coefficient for ditch conduit
w = unit weight of fill, lb/ft³
h = height of fill above top of conduit, ft
b = width of ditch at top of conduit, ft
D = outside diameter of conduit, ft

From the equilibrium of vertical forces, including shears, acting on the backfill above the conduit, C_D may be determined:

$$C_D = \frac{1 - e^{-kh/b}}{k} \frac{b}{h} \qquad (11.4)$$

where $e = 2.718$
$k = 2K_a \tan\theta$
K_a = coefficient of active earth pressure Table 11.2N
θ = angle of friction between fill and adjacent soil ($\theta \leq \phi$, angle of internal friction of fill)

Table 11.1N gives values of C_D for $k = 0.33$ for cohesionless soils, $k = 0.30$ for saturated topsoil, and $k = 0.26$ and 0.22 for clay (usual maximum and saturated).

Vertical load, lb/lin. ft, on conduit installed by tunneling may be estimated from

$$W = C_D b (wh - 2c) \qquad (11.5)$$

where c = cohesion of the soil, or half the unconfined compressive strength of the soil, psf. The load coefficient C_D may be computed from Eq. (11.4) or obtained from Table 11.1N with b = maximum width of tunnel excavation, ft, and h = distance from tunnel top to ground surface, ft.

For a ditch conduit, shearing forces extend from the pipe top to the ground surface. For a projecting conduit, however, if the embankment is sufficiently high, the shear may become zero at a horizontal plane below grade, the plane of equal settlement. Load on a projecting conduit is affected by the location of this plane.

Vertical load, lb/lin. ft, on a positive projecting conduit may be computed from

$$W = C_p whD \qquad (11.6)$$

where C_p = load coefficient for positive projecting conduit. Formulas have been derived for C_p and the depth of the plane of equal settlement. These formulas, however, are too lengthy for practical application, and the computation does not appear to be justified by the uncertainties in actual relative

Continued on page 214

TABLE 11.4 Pipes and Tunnels: Rectangular Cross Section

8.

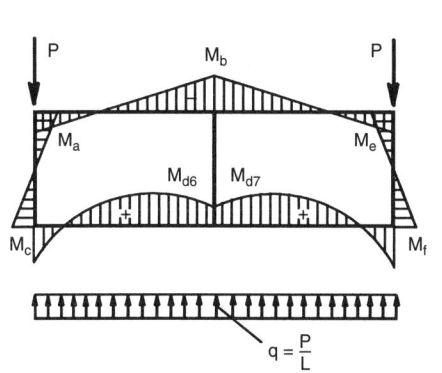

$$m_1 = 24(k+6)r$$

$$M_a = M_e = PL\frac{47k+18}{m_1}$$

$$M_{b1} = M_{b2} = -PL\frac{15k^2+49k+18}{m_1}$$

$$M_c = M_f = -PL\frac{49k+30}{m_1}$$

$$M_{d6} = M_{d7} = PL\frac{9k^2+11k+6}{m_1}$$

$$M_{b4} = M_{d4} = 0$$

9.

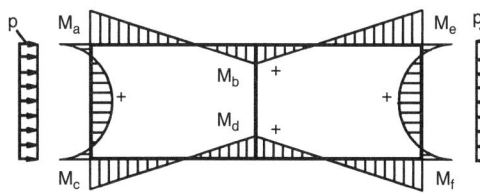

$$M_a = M_c = M_e = M_f = -\frac{ph^2}{6}\cdot\frac{k}{r}$$

$$M_{b1} = M_{b2} = M_{d7} = \frac{ph^2}{12}\cdot\frac{k}{r}$$

$$M_{b4} = M_{d4} = 0$$

10.

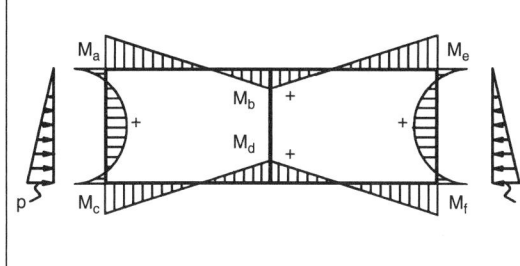

$$m_2 = \frac{20(k+6)r}{k}$$

$$M_a = M_e = -\frac{ph^2}{6}\cdot\frac{8k+59}{m_2}$$

$$M_c = M_f = -\frac{ph^2}{6}\cdot\frac{12k+61}{m_2}$$

$$M_{b1} = M_{b2} = \frac{ph^2}{6}\cdot\frac{7k+31}{m_2}$$

$$M_{d6} = M_{d7} = \frac{ph^2}{6}\cdot\frac{3k+29}{m_2}$$

$$M_{b4} = M_{d4} = 0$$

NOTES

settlement of the soil above the conduit (Fig. 11.2). Tests may be made in the field to determine C_p. If so, the possibility of an increase in earth pressure with time should be considered. For a rough estimate, C_p may be assumed as 1 for flexible conduit and 15 for rigid conduit.

The vertical load, lb/lin. ft, on negative projecting conduit may be computed from

$$W = C_N whb \tag{11.7}$$

where C_N = load coefficient for negative projecting conduit
h = height of fill above top of conduit, ft
b = horizontal width of trench at top of conduit, ft

The load on an imperfect ditch conduit (Fig. 11.3) may be obtained from

$$W = C_N whD \tag{11.8}$$

where D = outside diameter of conduit, ft.

Formulas have been derived for C_N, but they are complex, and insufficient values are available for the parameters involved. As a rough guide, C_N may be taken as 0.9 when depth of cover exceeds conduit diameter.

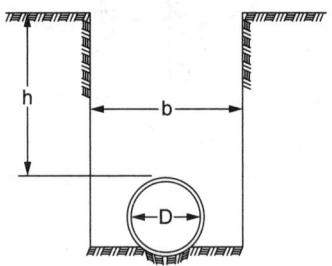

FIGURE 11.2 Ditch conduit.

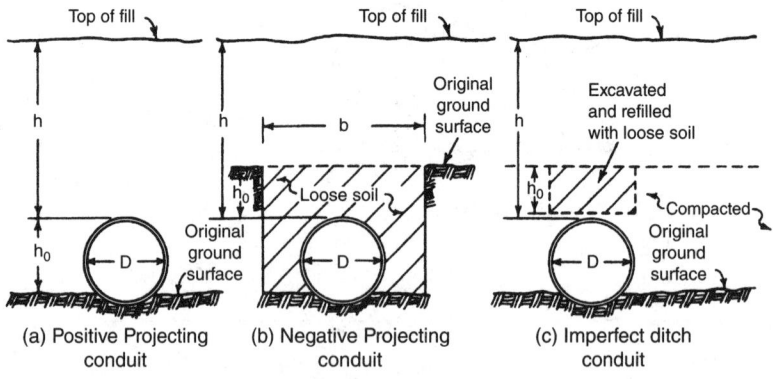

FIGURE 11.3 Type of projecting conduit depends on method of backfilling.

Continued on page 216

TABLE 11.5   Pipes and Tunnels: Rectangular Cross Section

11.

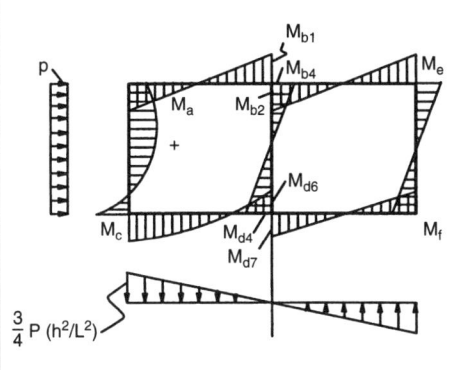

$\alpha_1 = 120k^3 + 278k^2 + 335k + 63$

$\alpha_2 = 360k^3 + 742k^2 + 285k + 27$

$\alpha_3 = 120k^3 + 529k^2 + 382k + 63$

$\alpha_4 = 120k^3 + 611k^2 + 558k + 87$

$$m = 20(k+2)(6k^2+6k+1), \quad n_1 = \frac{r}{k}$$

$$M_a = \frac{ph^2}{24}\left(-\frac{2}{n_1}+\frac{\alpha_1}{m}\right), \quad M_e = \frac{ph^2}{24}\left(-\frac{2}{n_1}-\frac{\alpha_1}{m}\right)$$

$$M_c = -\frac{ph^2}{24}\left(\frac{2}{n_1}+\frac{\alpha_2}{m}\right), \quad M_f = -\frac{ph^2}{24}\left(\frac{2}{n_1}-\frac{\alpha_2}{m}\right)$$

$$M_{b1} = -\frac{ph^2}{24}\left(-\frac{1}{n_1}+\frac{\alpha_3}{m}\right), \quad M_{b2} = -\frac{ph^2}{24}\left(-\frac{1}{n_1}-\frac{\alpha_3}{m}\right)$$

$$M_{d6} = -\frac{ph^2}{24}\left(-\frac{1}{n_1}+\frac{\alpha_4}{m}\right), \quad M_{d7} = -\frac{ph^2}{24}\left(-\frac{1}{n_1}-\frac{\alpha_4}{m}\right)$$

$$M_{b4} = -\frac{ph^2}{12}\cdot\frac{\alpha_3}{m}, \quad M_{d4} = \frac{ph^2}{12}\cdot\frac{\alpha_4}{m}$$

12.

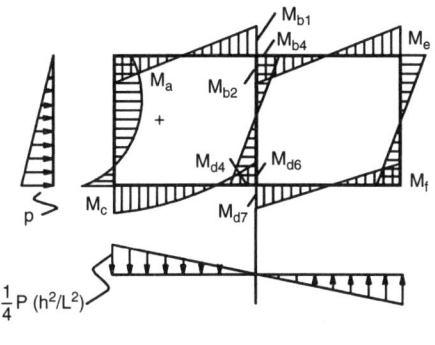

$\alpha_1 = 24k^3 + 50k^2 + 99k + 21$

$\alpha_2 = 144k^3 + 298k^2 + 109k + 9$

$\alpha_3 = 36k^3 + 169k^2 + 120k + 21$

$\alpha_4 = 36k^3 + 203k^2 + 192k + 29$

$$m = 20(k+2)(6k^2+6k+1), \quad n_2 = \frac{10(k+6)r}{k}$$

$$\frac{M_a}{M_e} = \frac{ph^2}{24}\left(-\frac{8k+59}{n_2}\pm\frac{\alpha_1}{m}\right)$$

$$\frac{M_c}{M_f} = -\frac{ph^2}{24}\left(\frac{12k+61}{n_2}\pm\frac{\alpha_2}{m}\right)$$

$$\frac{M_{b1}}{M_{b2}} = -\frac{ph^2}{24}\left(\frac{7k+31}{n_2}\pm\frac{\alpha_3}{m}\right)$$

$$\frac{M_{d6}}{M_{d7}} = \frac{ph^2}{24}\left(\frac{3k+29}{n_2}\pm\frac{\alpha_4}{m}\right)$$

$$M_{b4} = -\frac{ph^2}{12}\cdot\frac{\alpha_3}{m}, \quad M_{d4} = \frac{ph^2}{12}\cdot\frac{\alpha_4}{m}$$

─ **NOTES** ─

TABLE 11.1N Load Coefficients C_D for Ditch Conduit

h/b	Cohesionless Soils	Saturated Topsoil	Clay k = 0.26	Clay k = 0.22
1	0.85	0.86	0.88	0.89
2	0.75	0.75	0.78	0.80
3	0.63	0.67	0.69	0.73
4	0.55	0.58	0.62	0.67
5	0.50	0.52	0.56	0.60
6	0.44	0.47	0.51	0.55
7	0.39	0.42	0.46	0.51
8	0.35	0.38	0.42	0.47
9	0.32	0.34	0.39	0.43
10	0.30	0.32	0.36	0.40
11	0.27	0.29	0.33	0.37
12	0.25	0.27	0.31	0.35
Over 12	3.0b/h	3.3b/h	3.9b/h	4.5b/h

Coulomb derived the trigonometric equivalent:

$$k_a = \tan^2\left(45° - \frac{\phi}{2}\right) \tag{11.9}$$

TABLE 11.2N Active Lateral-Pressure Coefficients K_a

$\phi =$		10°	15°	20°	25°	30°	35°	40°
$\beta = 0$	$\alpha = 0$	0.70	0.59	0.49	0.41	0.33	0.27	0.22
	$\alpha = 10°$	0.97	0.70	0.57	0.47	0.37	0.30	0.24
	$\alpha = 20°$	—	—	0.88	0.57	0.44	0.34	0.27
	$\alpha = 30°$	—	—	—	—	0.75	0.43	0.32
	$\alpha = \phi$	0.97	0.93	0.88	0.82	0.75	0.67	0.59
$\beta = 10°$	$\alpha = 0$	0.76	0.65	0.55	0.48	0.41	0.43	0.29
	$\alpha = 10°$	1.05	0.78	0.64	0.55	0.47	0.38	0.32
	$\alpha = 20°$	—	—	1.02	0.69	0.55	0.45	0.36
	$\alpha = 30°$	—	—	—	—	0.92	0.56	0.43
	$\alpha = \phi$	1.05	1.04	1.02	0.98	0.92	0.86	0.79
$\beta = 20°$	$\alpha = 0$	0.83	0.74	0.65	0.57	0.50	0.43	0.38
	$\alpha = 10°$	1.17	0.90	0.77	0.66	0.57	0.49	0.43
	$\alpha = 20°$	—	—	1.21	0.83	0.69	0.57	0.49
	$\alpha = 30°$	—	—	—	—	1.17	0.73	0.59
	$\alpha = \phi$	1.17	1.20	1.21	1.20	1.17	1.12	1.06
$\beta = 30°$	$\alpha = 0$	0.94	0.86	0.78	0.70	0.62	0.56	0.49
	$\alpha = 10°$	1.37	1.06	0.94	0.83	0.74	0.65	0.56
	$\alpha = 20°$	—	—	1.51	1.06	0.89	0.77	0.66
	$\alpha = 30°$	—	—	—	—	1.55	0.99	0.79
	$\alpha = \phi$	1.37	1.45	1.51	1.54	1.55	1.54	1.51

Continued on page 218

TABLE 11.6 Pipes and Tunnels: Circular Cross Section

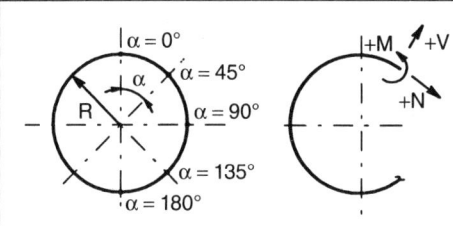

+M = tension on inside of ring
+ Tension
− Compression

Loading Condition			α = 0°	α = 45°	α = 90°	α = 135°	α = 180°
1.		M	+0.25wR²	0	−0.25wR²	0	+0.25wR²
		N	0	−0.5wR	−1.0wR	−0.5wR	0
		V	0	−0.5wR	0	+0.5wR	0
2.		M	−0.25pR²	0	+0.25pR²	0	−0.25pR²
		N	−1.0pR	−0.5pR	0	−0.5pR	−1.0pR
		V	0	+0.5pR	0	−0.5pR	0
3.		M	−0.208pR³	−0.029pR³	+0.25pR³	+0.029pR³	−0.292pR³
		N	−0.625pR²	−0.412pR²	0	−0.588pR²	−1.375pR²
		V	0	+0.411pR²	+0.125pR²	−0.589pR²	0
4.		M	0	0	0	0	0
		N	−pR	−pR	−pR	−pR	−pR
		V	0	0	0	0	0

NOTES

TABLE 11.3N Angles of Internal Friction and Unit Weights of Soils

Type of Soil	Density or Consistency	Angle of Internal Friction ϕ, deg	Unit Weight w, lb/ft³
Coarse sand or sand and gravel	Compact	40	140
	Loose	35	90
Medium sand	Compact	40	130
	Loose	30	90
Fine silty sand or sandy silt	Compact	30	130
	Loose	25	85
Uniform silt	Compact	30	135
	Loose	25	85
Clay-silt	Soft to medium	20	90–120
Silty clay	Soft to medium	15	90–120
Clay	Soft to medium	0–10	90–120

Pipe Stresses Perpendicular to the Longitudinal Axis

The stresses acting perpendicular to the longitudinal axis of a pipe are caused by either internal or external pressures on the pipe walls.

Internal pressure creates a stress commonly called hoop tension. It may be calculated by taking a free-body diagram of a 1-in (25.4-mm) long strip of pipe cut by a vertical plane through the longitudinal axis (Fig. 11.4). The forces in the vertical direction cancel out. The sum of the forces in the horizontal direction is

$$pD = 2F \qquad (11.10)$$

where P = internal pressure, lb/in² (MPa)
D = outside diameter of pipe, in (mm)
F = force acting on each cut of edge of pipe, lb (N)

Hence, the stress lb/in² (MPa), on the pipe material is

$$f = \frac{F}{A} = \frac{pD}{2t} \qquad (11.11)$$

where A = area of cut edge of pipe, ft² (m²), and t = thickness of pipe wall, in (mm).

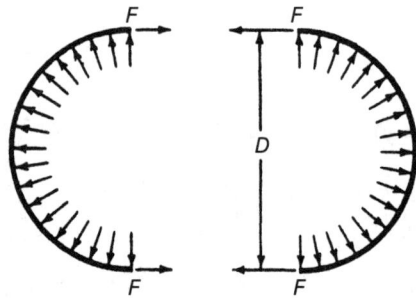

FIGURE 11.4 Internal pipe pressure produces hoop tension.

TABLE 11.7 Pipes and Tunnels: Circular Cross Section

Loading Condition			$\alpha = 0°$	$\alpha = 45°$	$\alpha = 90°$	$\alpha = 135°$	$\alpha = 180°$
5. (Ry, 0.2145 γR)		M	$+0.027\gamma R^3$	$+0.010\gamma R^3$	$-0.042\gamma R^3$	$-0.003\gamma R^3$	$+0.045\gamma R^3$
		N	$+0.021\gamma R^2$	$-0.030\gamma R^2$	$-0.215\gamma R^2$	$-0.122\gamma R^2$	$-0.021\gamma R^2$
		V	0	$-0.061\gamma R^2$	$-0.021\gamma R^2$	$+0.092\gamma R^2$	0
6. Buoyancy Forces $\gamma_w R(1-\cos\alpha)$		M	0	0	0	0	0
		N	$-0.5\gamma_w R^2$	$+0.646\gamma_w R^2$	$-1.0\gamma_w R^2$	$-1.354\gamma_w R^2$	$-1.5\gamma_w R^2$
		V	0	0	0	0	0
7. (45°, 30°, 30°, F, F)		M	$+0.151\gamma_w R^3$	$+0.026\gamma_w R^3$	$-0.176\gamma_w R^3$	$+0.001\gamma_w R^3$	$+0.121\gamma_w R^3$
		N	$-0.481\gamma_w R^2$	$+0.188\gamma_w R^2$	$+0.066\gamma_w R^2$	$+0.316\gamma_w R^2$	$+1.077\gamma_w R^2$
		V	0	$+0.191\gamma_w R^2$	$+0.016\gamma_w R^2$	$-0.567\gamma_w R^2$	0
8. (30°, 30°, F, F)		M	$+0.320\gamma_w R^3$	$+0.152\gamma_w R^3$	$-0.091\gamma_w R^3$	$+0.128\gamma_w R^3$	$+0.279\gamma_w R^3$
		N	$-0.821\gamma_w R^2$	$-0.653\gamma_w R^2$	$+0.090\gamma_w R^2$	$+1.366\gamma_w R^2$	$+1.59\gamma_w R^2$
		V	0	$+0.366\gamma_w R^2$	$+0.125\gamma_w R^2$	$-0.744\gamma_w R^2$	0

γ and γ_w = unit weight of soil and liquid, respectively

── N O T E S ──

Forces due to Pipe Bends

It is a common practice to use thrust blocks in pipe bends to take the forces on the pipe caused by the momentum change and the unbalanced internal pressure of the water.

The force diagram in Fig. 11.5 is a convenient method for finding the resultant force on a bend. The forces can be resolved into X and Y components to find the magnitude and direction of the resultant force on the pipe. In Fig. 11.5,

V_1 = velocity before change in size of pipe, ft/s (m/s)
V_2 = velocity after change in size of pipe, ft/s (m/s)
P_1 = pressure before bend or size change in pipe, lb/ft² (kPa)
P_2 = pressure after bend or size change in pipe, lb/ft² (kPa)
A_1 = area before size change in pipe, ft² (m²)
A_2 = area after size change in pipe, ft² (m²)
F_{1m} = force due to momentum of water in section 1 = $V_1 Qw/g$
P_1 = pressure of water in section 1 times area of section 1 = $P_1 A_1$
F_{2m} = force due to momentum of water in section 2 = $V_2 Qw/g$
P_2 = pressure of water in section 2 times area of section 2 = $P_2 A_2$
w = unit weight of liquid, lb/ft³ (kg/m³)
Q = discharge, ft³/s (m³/s)

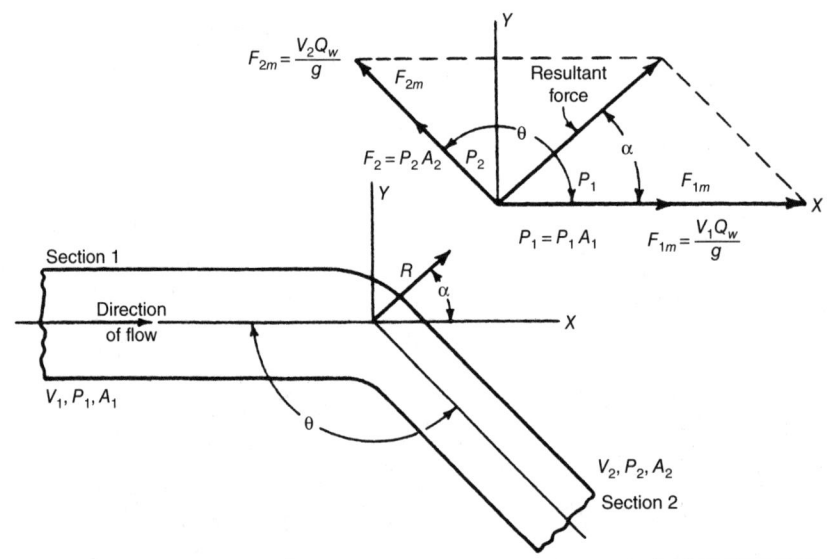

FIGURE 11.5 Forces produced by flow at a pipe bend and a change in diameter.

Continued on page 222

Pipes and Tunnels

TABLE 11.8 Pipes and Tunnels: Circular Cross Section

Loading Condition		$\alpha = 0°$	$\alpha = 45°$	$\alpha = 90°$	$\alpha = 135°$	$\alpha = 180°$
9.	M	$+0.378pR^2$	$+0.043pR^2$	$-0.442pR^2$	$-0.007pR^2$	$+0.308pR^2$
	N	$+0.25pR$	$-0.378pR$	$-1.570pR$	$-1.842pR$	$-0.25pR$
	V	0	$-0.732pR$	$+0.25pR$	$-1.488pR$	0
10.	M	$-0.137PR$	$-0.043PR$	$+0.182PR$	$+0.114PR$	$-0.500PR$
	N	$-0.318P$	$-0.225P$	$+1.0P$	$+0.939P$	$+0.318P$
	V	0	$-0.225P$	$-0.318P$	$+0.482P$	$+1.0P$
11.	M	$+0.318PR$	$+0.035PR$	$-0.182PR$	$+0.035PR$	$+0.318PR$
	N	0	$-0.354P$	$-0.5P$	$-0.354P$	0
	V	$+0.5P$	$+0.354P$	0	$-0.354P$	$-0.5P$
12.		$M_{max} = \dfrac{wR^2}{4} - \dfrac{R^2}{48}(5p_1 + 7p_2)$				
		$M_{min} = -\dfrac{wR^2}{4} + \dfrac{R^2}{8}(p_1 + p_2)$				
		$N = \dfrac{R(11p_1 + 5p_2)}{16}$				
		If $p_1 = p_2 = p$: $M_{max} = \dfrac{R^2}{4}(w - p)$, $M_{min} = -\dfrac{R^2}{4}(w - p)$				
		$N = pR$				

─── NOTES ───

If the pressure loss in the bend is neglected and there is no change in magnitude of velocity around the bend, a quick solution is

$$R = 2A\left(w\frac{V^2}{g} + p\right)\cos\frac{\theta}{2} \qquad (11.12)$$

$$\alpha = \frac{\theta}{2}$$

where R = resultant force on bend, lb (N)
 α = angle R makes with F_{1m}
 p = pressure, lb/ft² (kPa)
 w = unit weight of water, 62.4 lb/ft³ (998.4 kg/m³)
 V = velocity of flow, ft/s (m/s)
 g = acceleration due to gravity, 32.2 ft/s² (9.81 m/s²)
 A = area of pipe, ft² (m²)
 θ = angle between pipes ($0° \le \theta \le 180°$)

Pipe on Supports at Intervals

1. For a circular pipe or tank supported at intervals and held circular at the supports by rings or bulkheads, the ordinary theory of flexure is applicable if the pipe is completely filled.
2. If the pipe is only partially filled, the cross section at points between supports becomes out of round, and the distribution of longitudinal fiber stress is neither linear nor symmetrical across the section. The highest stresses occur for the half-full condition; then the maximum longitudinal compressive stress and the maximum circumferential bending stresses occur at the ends of the horizontal diameter; the maximum longitudinal tensile stress occurs at the bottom; and the longitudinal stress at the top is practically zero. According to theory (Ref. 1), the greatest of these stresses is the longitudinal compression, which is equal to the maximum longitudinal stress for the full condition divided by $k^{\frac{1}{2}} = \left(\frac{L}{R}\sqrt{\frac{t}{R}}\right)^{\frac{1}{2}}$, where R = the pipe radius, t = the thickness, and L = the span. The maximum circumferential stress is about one-third of this. Tests (Ref. 2) on a pipe having $K = 1.36$ showed a longitudinal stress, somewhat less, and a circumferential stress considerably greater, than indicated by this theory.

---

[1] H. Schorer, Design of Large Pipe Lines, *Trans. Am. Soc. Civil Eng.*, vol. 98, p. 101, 1933.
[2] R. S. Hatenberg, The Strength and Stiffness of Thin Cylinderi Shells on Saddle Supports, doctor a dissertation, University of Wisconsin, 1941.

NOTES

3. For an unstiffened pipe resting is saddle supports, there are high local stresses both longitudinal and circumferential, adjacent to the tips of the saddles. These stresses are less for a large saddle angle β (total angle subtended by arc of contact between pipe and saddle) than for a small angle and for the ordinary range of dimensions are practically independent of the thickness of the saddle, i.e., its dimension parallel to the pipe axis. The maximum value of these localized stresses, for a pipe that fits the saddle well, will probably not exceed that indicated by the formula

$$\delta_{max} = k \frac{P}{t^2} \log_e \left(\frac{R}{t} \right)$$

where P = the total saddle reaction, R = the pipe radius, i = pipe thickness, and k = a coefficient given by

$$k = 0.02 - 0.00012(\beta - 90)$$

where β is in degrees. This stress is almost wholly due to circumferential bending, and occurs at points about 15° above the saddle tips.

4. The maximum value of P the pipe can sustain is about 2.25 times the value that, according to the formula given above, will produce a maximum stress equal to the yield point of the pipe material.

5. For a pipe supported in flexible slings instead of on rigid saddles, the maximum local stresses occur at the points of tangency of sling an pipe section; they are in general less than the corresponding stresses in the saddle supported pipe, but are of the same order of magnitude.

APPENDIX A
Quick-Use Conversion Tables

Structural engineers and designers work on projects throughout the world using both SI (Système International) and the USCS (United States Customary System). To allow easy conversion between such systems, Tables A.1, A.2, A.3, and A.4 have been prepared.

Table A.1 is a working structural specialist's listing of the units he or she frequently uses in daily work. It allows fast conversion without having to refer to a longer table. Any working structural specialist can construct such a table using Table A.2 as the source of conversion factors, which are simple multipliers. Table A.2 is a comprehensive listing of conversion factors for a variety of engineering activities.

Tables A.3 and A.4 are the work of Dr. Mikhelson, based on his wide experience in structural engineering. Both tables are highly useful in actual on-the-job design or erection work.

Appendix A

TABLE A.1 Typical Conversion Table[†]

To Convert from	To	Multiply by[‡]	
square foot	square meter	9.290304	E – 02
foot per second squared	meter per second squared	3.048	E – 01
cubic foot	cubic meter	2.831685	E – 02
pound per cubic inch	kilogram per cubic meter	2.767990	E + 04
gallon per minute	liter per second	6.309	E – 02
pound per square inch	kilopascal	6.894757	
pound force	newton	4.448222	
kip per square foot	pascal	4.788026	E + 04
acre foot per day	cubic meter per second	1.427641	E – 02
acre	square meter	4.046873	E + 03
cubic foot per second	cubic meter per second	2.831685	E – 02

[†]This table contains only selected values. See the U.S. Department of the Interior, *Metric Manual*, or National Bureau of Standards, *The International System of Units* (SI), both available from the U.S. Government Printing Office (GPO), for far more comprehensive listings of conversion factors.

[‡]The E indicates an exponent, as in scientific notation, followed by a positive or negative number, representing the power of 10 by which the given conversion factor is to be multiplied before use. Thus, for the square foot conversion factor, $9.290304 \times 1/100 = 0.09290304$, the factor to be used to convert square feet to square meters. For a positive exponent, as in converting acres to square meters, multiply by $4.046873 \times 1000 = 4046.8$.

Where a conversion factor cannot be found, simply use the dimensional substitution. Thus, to convert pounds per cubic inch to kilograms per cubic meter, find 1 lb = 0.4535924 kg and 1 in^3 = 0.00001638706 m^3. Then 1 lb/in^3 = 0.4535924 kg/0.00001638706 m^3 = 27,680.01, or 2.768 E + 4.

Conversion Factors

TABLE A.2 Factors for Conversion to SI Units of Measurement

To Convert from	To	Multiply by	
acre	square meter, m^2	4.046873	E + 03
acre foot, acre ft	cubic meter, m^3	1.233489	E + 03
angstrom, Å	meter, m	1.000000*	E − 10
atmosphere, atm (standard)	pascal, Pa	1.013250*	E + 05
atmosphere, atm (technical = 1 kgf/cm^2)	pascal, Pa	9.806650*	E + 04
bar	pascal, Pa	1.000000*	E + 05
barrel (for petroleum, 42 gal)	cubic meter, m^2	1.589873	E − 01
board foot, board ft	cubic meter, m^3	2.359737	E − 03
British thermal unit, Btu (mean)	joule, J	1.05587	E + 03
British thermal unit, Btu (International Table) · in/(h)(ft^2) (°F) (k, thermal conductivity)	watt per meter kelvin, W/(m · k)	1.442279	E − 01
British thermal unit, Btu (International Table)/h	watt, W	2.930711	E − 01
British thermal unit, Btu (International Table)/(h)(ft^2)(°F) (C, thermal conductance)	watt per square meter kelvin, W/(m^2·K)	5.678263	E + 00
British thermal unit, Btu (International Table)/lb	joule per kilogram, J/kg	2.326000*	E + 03
British thermal unit, Btu (International Table)/(lb)(°F) (C, heat capacity)	joule per kilogram kelvin, J/(kg·K)	4.186800*	E + 03
British thermal unit, cubic foot, Btu (International Table)/ft^3	joule per cubic meter, J/m^3	3.725895	E + 04
bushel (U.S.)	cubic meter, m^3	3.523907	E − 02
calorie (mean)	joule, J	4.19002	E + 00
candela per square inch, cd/in^2	candela per square meter, cd/m^2	1.550003	E + 03
centimeter, cm, of mercury (0°C)	pascal, Pa	1.33322	E + 03
centimeter, cm, of water (4°C)	pascal, Pa	9.80638	E + 01
chain	meter, m	2.011684	E + 01
circular mil	square meter, m^2	5.067075	E − 10
day	second, s	8.640000*	E + 04
day (sidereal)	second, s	8.616409	E + 04
degree (angle)	radian, rad	1.745329	E − 02

Appendix A

TABLE A.2 Factors for Conversion to SI Units of Measurement (*Continued*)

To Convert from	To	Multiply by	
degree Celsius	kelvin, K	$T_K = t_C + 273.15$	
degree Fahrenheit	degree Celsius, °C	$t_C = (t_F - 32)/1.8$	
degree Fahrenheit	kelvin, K	$T_K = (t_F + 459.67)/1.8$	
degree Rankine	kelvin, K	$T_K = T_R/1.8$	
(°F)(h)(ft²)/Btu (International Table) (R, thermal resistance)	kelvin square meter per watt, K · m²/W	1.761102	E − 01
(°F)(h)(ft²)/(Btu (International Table) · in) (thermal resistivity)	kelvin meter per watt, K · m/W	6.933471	E + 00
dyne, dyn	newton, N	1.000000†	E − 05
fathom	meter, m	1.828804	E + 00
foot, ft	meter, m	3.048000†	E − 01
foot, ft (U.S. survey)	meter, m	3.048006	E − 01
foot, ft, of water (39.2°F) (pressure)	pascal, Pa	2.98898	E + 03
square foot, ft²	square meter, m²	9.290304†	E − 02
square foot per hour, ft²/h (thermal diffusivity)	square meter per second, m²/s	2.580640†	E − 05
square foot per second, ft²/s	square meter per second, m²/s	9.290304†	E − 02
cubic foot, ft³ (volume or section modulus)	cubic meter, m³	2.831685	E − 02
cubic foot per minute, ft³/min	cubic meter per second, m³/s	4.719474	E − 04
cubic foot per second, ft³/s	cubic meter per second, m³/s	2.831685	E − 02
foot to the fourth power, ft⁴ (area moment of inertia)	meter to the fourth power, m⁴	8.630975	E − 03
foot per minute, ft/min	meter per second, m/s	5.080000†	E − 03
foot per second, ft/s	meter per second, m/s	3.048000†	E − 01
foot per second squared, ft/s²	meter per second squared, m/s²	3.048000†	E − 01
footcandle, fc	lux, lx	1.076391	E + 01
footlambert, fL	candela per square meter, cd/m²	3.426259	E + 00
foot pound force, ft · lbf	joule, J	1.355818	E + 00
foot pound force per minute, ft · lbf/min	watt, W	2.259697	E − 02
foot pound force per second, ft · lbf/s	watt, W	1.355818	E + 00
foot poundal, ft poundal	joule, J	4.214011	E − 02

TABLE A.2 Factors for Conversion to SI Units of Measurement (*Continued*)

To Convert from	To	Multiply by	
free fall, standard g	meter per second squared, m/s^2	9.806650†	E + 00
gallon, gal (Canadian liquid)	cubic meter, m^3	4.546092	E – 03
gallon, gal (U.K. liquid)	cubic meter, m^3	4.546092	E – 03
gallon, gal (U.S. dry)	cubic meter, m^3	4.404884	E – 03
gallon, gal (U.S. liquid)	cubic meter, m^3	3.785412	E – 03
gallon, gal (U.S. liquid) per day	cubic meter per second, m^3/s	4.381264	E – 08
gallon, gal (U.S. liquid) per minute	cubic meter per second, m^3/s	6.309020	E – 05
grad	degree (angular)	9.000000†	E – 01
grad	radian, rad	1.570796	E – 02
grain, gr	kilogram, kg	6.479891†	E – 05
gram, g	kilogram, kg	1.000000†	E – 03
hectare, ha	square meter, m^2	1.000000†	E + 04
horsepower, hp (550 ft · lbf/s)	watt, W	7.456999	E + 02
horsepower, hp (boiler)	watt, W	9.80950	E + 03
horsepower, hp (electric)	watt, W	7.460000†	E + 02
horsepower, hp (water)	watt, W	7.46043†	E + 02
horsepower, hp (U.K.)	watt, W	7.4570	E + 02
hour, h	second, s	3.600000†	E + 03
hour, h (sidereal)	second, s	3.590170	E + 03
inch, in	meter, m	2.540000†	E – 02
inch of mercury, in Hg (32°F) (pressure)	pascal, Pa	3.38638	E + 03
inch of mercury, in Hg (60°F) (pressure)	pascal, Pa	3.37685	E + 03
inch of water, in H$_2$O (60°F) (pressure)	pascal, Pa	2.4884	E + 02
square inch, in^2	square meter, m^2	6.451600†	E – 04
cubic inch, in^3 (volume or section modulus)	cubic meter, m^3	1.638706	E – 05
inch to the fourth power, in^4 (area moment of inertia)	meter to the fourth power, m^4	4.162314	E – 07
inch per second, in/s	meter per second, m/s	2.540000†	E – 02
kelvin, K	degree Celsius, °C	$t_C = T_K - 273.15$	
kilogram force, kgf	newton, N	9.806650†	E + 00
kilogram force meter, kg · m	newton meter, N · m	9.806650†	E + 00

TABLE A.2 Factors for Conversion to SI Units of Measurement (*Continued*)

To Convert from	To	Multiply by	
kilogram force second squared per meter, kgf · s²/m (mass)	kilogram, kg	9.806650†	E + 00
kilogram force per square centimeter, kgf/cm²	pascal, Pa	9.806650†	E + 04
kilogram force per square meter, kgf/m²	pascal, Pa	9.806650†	E + 00
kilogram force per square millimeter, kgf/mm²	pascal, Pa	9.806650†	E + 06
kilometer per hour, km/h	meter per second, m/s	2.777778	E − 01
kilowatt hour, kWh	joule, J	3.600000†	E + 06
kip (1000 lbf)	newton, N	4.448222	E + 03
kip per square inch, kip/in², ksi	pascal, Pa	6.894757	E + 06
knot, kn (international)	meter per second, m/s	5.144444	E − 01
lambert, L	candela per square meter, cd/m²	3.183099	E + 03
liter	cubic meter, m³	1.000000†	E − 03
maxwell	weber, Wb	1.000000†	E − 08
mho	siemens, S	1.000000†	E + 00
microinch, μin	meter, m	2.540000†	E − 08
micrometer, micron, μm	meter, m	1.000000†	E − 06
mil, mi	meter, m	2.540000†	E − 05
mile, mi (international)	meter, m	1.609344†	E + 03
mile, mi (U.S. statute)	meter, m	1.609347	E + 03
mile, mi (international nautical)	meter, m	1.852000†	E + 03
mile, mi (U.S. nautical)	meter, m	1.852000†	E + 03
square mile, mi² (international)	square meter, m²	2.589988	E + 06
square mile, mi² (U.S. statute)	square meter, m²	2.589998	E + 06
mile per hour, mi/h (international)	meter per second, m/s	4.470400†	E − 01
mile per hour, mi/h (international)	kilometer per hour, km/h	1.609344†	E + 00
millibar, mbar	pascal, Pa	1.000000†	E + 02
millimeter of mercury, mmHg (0°C)	pascal, Pa	1.33322	E + 02
minute (angle)	radian, rad	2.908882	E − 04
minute min	second, s	6.000000†	E + 01
minute (sidereal)	second, s	5.983617	E + 01
ounce, oz (avoirdupois)	kilogram, kg	2.834952	E − 02
ounce, oz (troy or apothecary)	kilogram, kg	3.110348	E − 02
ounce, oz (U.K. fluid)	cubic meter, m³	2.841307	E − 05

TABLE A.2 Factors for Conversion to SI Units of Measurement (*Continued*)

To Convert from	To	Multiply by	
ounce, oz (U.S. fluid)	cubic meter, m^3	2.957353	E – 05
ounce force, ozf	newton, N	2.780139	E – 01
ounce force · inch, ozf · in	newton meter, N · m	7.061552	E – 03
ounce per square foot, oz (avoirdupois)/ft^2	kilogram per square meter, kg/m^2	3.051517	E – 01
ounce per square yard, oz (avoirdupois)/yd^2	kilogram per square meter, kg/m^2	3.390575	E – 02
perm (0°C)	kilogram per pascal second meter, kg/(Pa · s · m)	5.72135	E – 11
perm (23°C)	kilogram per pascal second meter, kg/(Pa · s · m)	5.74525	E – 11
perm inch, perm · in (0°C)	kilogram per pascal second meter, kg/(Pa · s · m)	1.45322	E – 12
perm inch, perm · in (23°C)	kilogram per pascal second meter, kg/(Pa · s · m)	1.45929	E – 12
pint, pt (U.S. dry)	cubic meter, m^3	5.506105	E – 04
pint, pt (U.S. liquid)	cubic meter, m^3	4.731765	E – 04
poise, P (absolute viscosity)	pascal second, Pa · s	1.000000†	E – 01
pound, lb (avoirdupois)	kilogram, kg	4.535924	E – 01
pound, lb (troy or apothecary)	kilogram, kg	3.732417	E – 01
pound square inch, lb · in^2 (moment of inertia)	kilogram square meter, kg · m^2	2.926397	E – 04
pound per foot second, lb/ft · s	pascal second, Pa · s	1.488164	E + 00
pound per square foot, lb/ft^2	kilogram per square meter, kg/m^2	4.882428	E + 00
pound per cubic foot, lb/ft^3	kilogram per cubic meter, kg/m^3	1.601846	E – 01
pound per gallon, lb/gal (U.K. liquid)	kilogram per cubic meter, kg/m^3	9.977633	E + 01
pound per gallon, lb/gal (U.S. liquid)	kilogram per cubic meter, kg/m^3	1.198264	E + 02
pound per hour, lb/h	kilogram per second, kg/s	1.259979	E – 04
pound per cubic inch, lb/in^3	kilogram per cubic meter, kg/m^3	2.767990	E + 04
pound per minute, lb/min	kilogram per second, kg/s	7.559873	E – 03
pound per second, lb/s	kilogram per second, kg/s	4.535924	E – 01
pound per cubic yard, lb/yd^3	kilogram per cubic meter, kg/m^3	5.932764	E – 01
poundal	newton, N	1.382550	E – 01
pound force, lbf	newton, N	4.448222	E + 00

TABLE A.2 Factors for Conversion to SI Units of Measurement (*Continued*)

To Convert from	To	Multiply by	
pound force foot, lbf · ft	newton meter, N · m	1.355818	E + 00
pound force per foot, lbf/ft	newton per meter, N/m	1.459390	E + 01
pound force per square foot, lbf/ft^2	pascal, Pa	4.788026	E + 01
pound force per inch, lbf/in	newton per meter, N/m	1.751268	E + 02
pound force per square inch, lbf/in^2 (psi)	pascal, Pa	6.894757	E + 03
quart, qt (U.S. dry)	cubic meter, m^3	1.101221	E − 03
quart, qt (U.S. liquid)	cubic meter, m^3	9.463529	E − 04
rod	meter, m	5.029210	E + 00
second (angle)	radian, rad	4.848137	E − 06
second (sidereal)	second, s	9.972696	E − 01
square (100 ft^2)	square meter, m^2	9.290304†	E + 00
ton (assay)	kilogram, kg	2.916667	E − 02
ton (long, 2240 lb)	kilogram, kg	1.016047	E + 03
ton (metric)	kilogram, kg	1.000000†	E + 03
ton (refrigeration)	watt, W	3.516800	E + 03
ton (register)	cubic meter, m^3	2.831685	E + 00
ton (short, 2000 lb)	kilogram, kg	9.071847	E + 02
ton (long per cubic yard, ton)/yd^3	kilogram per cubic meter, kg/m^3	1.328939	E + 03
ton (short per cubic yard, ton)/yd^3	kilogram per cubic meter, kg/m^3	1.186553	E + 03
ton force (2000 lbf)	newton, N	8.896444	E + 03
tonne, t	kilogram, kg	1.000000†	E + 03
watt hour, Wh	joule, J	3.600000†	E + 03
yard, yd	meter, m	9.144000†	E − 01
square yard, yd^2	square meter, m^2	8.361274	E − 01
cubic yard, yd^3	cubic meter, m^3	7.645549	E − 01
year (365 days)	second, s	3.153600†	E + 07
year (sidereal)	second, s	3.155815	E + 07

†Exact value.
From E380, "Standard for Metric Practice," American Society for Testing and Materials.

TABLE A3 Conversion between Anglo-American and Metric Systems

Metric Units	Conversion Factors	
Units of Length		
1 millimeter (mm)	1 inch (in) = 25.4 mm	1 mm = 0.03937 in
1 centimeter (cm) = 10 mm	1 foot (ft) = 12 in = 304.8 mm	1 cm = 0.3937 in
1 decimeter (dm) = 10 cm = 100 mm	1 yard (yd) = 3 ft = 0.9144 m	1 m = 1.0904 yd
1 meter (m) = 100 cm = 1000 mm	1 mile (mi) = 1760 yd = 1609.344 m	1 km = 3281 ft
1 kilometer (km) = 1000 m	1 mi = 1.6093 km	1 km = 0.6214 mi
Units of Area		
1 square millimeter (mm^2)	1 square inch (in^2) = 645.16 mm^2	1 mm^2 = 0.001550 in^2
1 square centimeter (cm^2) = 100 mm^2	1 square foot (ft^2) = 0.092903 m^2	1 cm^2 = 0.1550 in^2
1 square meter (m^2) = 10^6 mm^2	1 square yard (yd^2) = 0.836127 m^2	1 m^2 = 10.76 ft^2
1 square kilometer (km^2) = 10^6 m^2	1 acre = 4046.856 m^2	1 m^2 = 1.19599 yd^2
1 hectare (ha) = 10^4 m^2 = 0.01 km^2	1 square mile (mi^2) = 2.5898 km^2	1 km^2 = 0.3861 mi^2
Units of Volume		
1 cubic millimeter (mm^3)	1 cubic inch (in^3) = 16387.064 mm^3	1 mm^3 = 0.00006102 in^3
1 cubic centimeter (cm^3) = 10^3 mm^3	1 cubic foot (ft^3) = 0.02831685 m^3	1 cm^3 = 0.06102 in^3
1 cubic meter (m^3) = 10^9 mm^3	1 cubic yard (yd^3) = 0.764555 m^3	1 m^3 = 1.30795 yd^3
1 cubic kilometer (km^3) = 10^9 m^3	1 acre · foot = 1233.482 m^3	1 m^3 = 35.31 ft^3
1 liter (L) = 1000 cm^3 = 0.001 m^3	1 gallon (gal) = 3.785412 L	1 L = 0.264172 gal
Units of Mass		
1 milligram (mg)	1 ounce (oz) = 28.34952 g	Mass per unit length
1 gram (g) = 1000 mg	1 pound (lb) = 0.453592 kg	1 lb/ft = 1.48816 kg/m
1 kilogram (kg) = 1000 g	1 kip = 453.592 kg	Mass per unit area
1 ton (t) = 1000 kg	1 ton (2000 lb) = 907.184 kg	1 lb/ft^2 = 4.88243 kg/m^2
		Mass per unit volume (mass density)
		1 lb/ft^3 = 16.01846 kg/m^3
		1 lb/yd^3 = 0.593276 kg/m^3

TABLE A3 Conversion between Anglo-American and Metric Systems (*Continued*)

Metric Units	Conversion Factors
Units of Force	
1 newton (N) = 1 kg (mass)/(m/s^2) 1 kilonewton (kN) = 1000 N 1 meganewton (MN) = 1000 kN Gravitational force: 1 N = 1 kg (mass)/9.81 = 0.102 kg or 1 kg (force) = 9.81 N Unit weight: 1 lb/ft^3 = 0.1571 kN/m^3	1 lb = 4.448222 N 1 kip = 4.448222 kN 1 ton (2000 lb) = 8.896444 kN 1 N = 0.2248 lb 1 kN = 0.2248 kip 1 kN = 0.1124 ton 1 kN/m^3 = 6.366 lb/ft^3	Force per unit length 1 lb/in = 175.1268 N/m 1 lb/ft = 14.5939 N/m Moment of force 1 lb·in = 0.112985 N·m 1 lb·ft = 1.355818 N·m

Metric Units	Conversion Factors	
Units of Pressure, Stress, Modulus of Elasticity		
1 pascal (Pa) = 1 N/m^2 1 kilopascal (kPa) = 1000 Pa = 1 kN/m^2 1 megapascal (MPa) = 1000 kPa 1 gigapascal (GPa) = 1000 MPa 1 atmosphere (atm) = 1 kg/cm^2 = 98.1 kPa 1 bar = 1.02 kg/cm^2 = 100 kPa	1 lb/in^2 = 6.894757 kPa 1 kip/in^2 = 6.894757 MPa 1 lb/ft^2 = 47.88026 Pa 1 kip/ft^2 = 47.88026 kPa 1 lb/in^2 = 0.07029 kg/cm^2	1 kPa = 0.145038 lb/in^2 1 MPa = 0.145038 kip/in^2 1 Pa = 0.020885 lb/ft^2 1 kPa = 0.020885 kip/ft^2 1 kg/cm^2 = 14.23 lb/in^2

Temperature:

$T_C^\circ = \frac{5}{9}(T_F^\circ - 32^\circ)$, where T_C° and T_F° are Celsius and Fahrenheit temperatures, respectively.

APPENDIX B
Mathematical Formulas: Algebra

TABLE B.1 Algebra

Powers		Roots	
$a^m \cdot a^n = a^{m+n}$	$\dfrac{a^m}{a^n} = a^{m-n}$	$a^{m/n} = \sqrt[n]{a^m}$	$\sqrt[m]{\sqrt[n]{a}} = \sqrt[m \cdot n]{a}$
$(a^m)^n = a^{m \cdot n}$	$(a \cdot b)^m = a^m \cdot b^m$	$\sqrt[m]{a} \cdot \sqrt[m]{b} = \sqrt[m]{a \cdot b}$	$\dfrac{\sqrt[m]{a}}{\sqrt[m]{b}} = \sqrt[m]{\dfrac{a}{b}}$
$\left(\dfrac{a}{b}\right)^m = \dfrac{a^m}{b^m}$	$a^m \cdot b \pm a^m \cdot c = (b \pm c)a^m$	$\left(\sqrt[m]{a}\right)^n = \sqrt[m]{a^n}$	$\sqrt[n]{\sqrt[m]{a}} = \sqrt[m \cdot n]{a}$
$a^{-m} = \dfrac{1}{a^m}$	$a^0 = 1$ when $a \neq 0$	$i = \sqrt{-1}$	$\sqrt{-a} = i \cdot \sqrt{a}$

Logarithms	$\log_a N = n$
	a = base, N = antilogarithm,
	n = logarithm (log)
	$\log_{10} = \log$ = common log,
	$\log_e = \ln$ = natural log
$\log_a(x \cdot y) = \log_a x + \log_a y$	$e = 2.718281828459 \cdots$
$\log_a\left(\dfrac{x}{y}\right) = \log_a x - \log_a y$	$\log 0.01 = -2$, $\log 0.1 = -1$, $\log 1 = 0$, $\log 10 = 1$, $\log 100 = 2$
$\log_a x^m = m \cdot \log_a x$	$\log x = \log e \cdot \ln x = 0.434294 \cdot \ln x$
$\log_a \sqrt[m]{x} = \dfrac{1}{m} \log_a x$	$\ln x = \dfrac{\log x}{\log e} = 2.302585 \log x$

Factorials	$n! = 1 \cdot 2 \cdot 3 \cdots n$
	$(n + 1)! = (n + 1)n!$
	$0! = 1$, $(0 + 1)! = (0 + 1)0!$
	$n! \approx \sqrt{2\pi n}\left(\dfrac{n}{e}\right)^n$

Permutations	Combinations
$P_m^n = \dfrac{n!}{(n-m)!} = n \cdot (n-1) \cdot (n-2) \cdots (n-m+1)$ $n \geq m$	$C_m^n = \dfrac{n!}{m!(n-m)!}$ $n \geq m$
Example: $P_3^5 = \dfrac{1 \cdot 2 \cdot 3 \cdot 4 \cdot 5}{1 \cdot 2} = 60$	**Example:** $C_3^5 = \dfrac{1 \cdot 2 \cdot 3 \cdot 4 \cdot 5}{1 \cdot 2 \cdot 3 \cdot (1 \cdot 2)} = 10$

where P = number of possible permutations, C = number of possible combinations, n = number of things given, m = number of selections from n given things.

TABLE B.1 Algebra (Continued)

Algebraic Expressions

$(a \pm b)^2 = a^2 \pm 2ab + b^2$	$a^2 - b^2 = (a+b)(a-b)$
$(a \pm b)^3 = a^3 \pm 3a^2b + 3ab^2 \pm b^3$	$a^3 \pm b^3 = (a \pm b)(a^2 \mp ab + b^2)$

$$(a+b)^n = a^n + \frac{n}{1}a^{n-1}b + \frac{n(n-1)}{1 \cdot 2}a^{n-2}b^2 + \frac{n(n-1)(n-2)}{1 \cdot 2 \cdot 3}a^{n-3}b^3 + \cdots + b^n$$

$$a^n - b^n = (a-b)(a^{n-1} + a^{n-2}b + a^{n-3}b^2 + \cdots + ab^{n-2} + b^{n-1})$$

Algebraic Equations

Linear Equations

Third-order determinants:
$$\begin{aligned} a_{11}x + a_{12}y + a_{13}z &= b_1 \\ a_{21}x + a_{22}y + a_{23}z &= b_2 \\ a_{31}x + a_{32}y + a_{33}z &= b_3 \end{aligned} \qquad x = \frac{D_1}{D}, \quad y = \frac{D_2}{D}, \quad z = \frac{D_3}{D}$$

$$D = \begin{vmatrix} a_{11} & a_{12} & a_{13} \\ a_{21} & a_{22} & a_{23} \\ a_{31} & a_{32} & a_{33} \end{vmatrix} = \begin{matrix} a_{11} \cdot a_{22} \cdot a_{33} - a_{11} \cdot a_{23} \cdot a_{32} \\ + a_{12} \cdot a_{23} \cdot a_{31} - a_{12} \cdot a_{21} \cdot a_{33} \\ + a_{13} \cdot a_{21} \cdot a_{32} - a_{13} \cdot a_{22} \cdot a_{31} \end{matrix} \qquad D_1 = \begin{vmatrix} b_1 & a_{12} & a_{13} \\ b_2 & a_{22} & a_{23} \\ b_3 & a_{32} & a_{33} \end{vmatrix} = \begin{matrix} b_1 \cdot a_{22} \cdot a_{33} - b_1 \cdot a_{23} \cdot a_{32} \\ + a_{12} \cdot a_{23} \cdot b_3 - a_{12} \cdot b_2 \cdot a_{33} \\ + a_{13} \cdot b_2 \cdot a_{32} - a_{13} \cdot a_{22} \cdot b_3 \end{matrix}$$

Determine D_2 and D_3 similarly by replacing the y and z columns by the b column.

Equation of the Second Degree

$x^2 + px + q = 0$	$x_{1,2} = -\dfrac{p}{2} \pm \sqrt{\left(\dfrac{p}{2}\right)^2 - q}$

Equation of the Third Degree

	$x_1 = y_1 - \dfrac{a}{3}$	Determinant: $D = \left(\dfrac{p}{3}\right)^3 + \left(\dfrac{q}{2}\right)^2$, $\quad p = b - \dfrac{a^3}{3}$, $\quad q = \dfrac{2}{27}a^3 - \dfrac{1}{3}a \cdot b + c$
$x^3 + ax^2 + bx + c = 0$	$x_2 = y_2 - \dfrac{a}{3}$	If $D = 0$: $y_1 = \sqrt[3]{-4q}$, $\quad y_2 = y_3 = \sqrt[3]{\dfrac{q}{2}}$
	$x_3 = y_3 - \dfrac{a}{3}$	If $D > 0$: $\omega_1 = \dfrac{-1 + i\sqrt{3}}{2}$, $\quad \omega_2 = \dfrac{-1 - i\sqrt{3}}{2}$

$$y_1 = \sqrt[3]{-\frac{q}{2} + \sqrt{D}} + \sqrt[3]{-\frac{q}{2} - \sqrt{D}}, \quad y_2 = \omega_1 \sqrt[3]{-\frac{q}{2} + \sqrt{D}} + \omega_2 \sqrt[3]{-\frac{q}{2} - \sqrt{D}},$$

$$y_3 = \omega_2 \sqrt[3]{-\frac{q}{2} + \sqrt{D}} + \omega_1 \sqrt[3]{-\frac{q}{2} - \sqrt{D}}$$

If $D < 0$: $y_1 = \dfrac{2}{3}\sqrt{3}\sqrt{|p|}\cos\phi$, $\quad y_2 = \dfrac{2}{3}\sqrt{3}\sqrt{|p|}\cos(\phi + 120°)$,

$\qquad\qquad y_3 = \dfrac{2}{3}\sqrt{3}\sqrt{|p|}\cos(\phi - 120°) \quad \phi = \dfrac{1}{3}\arccos\dfrac{-3\sqrt{3q}}{2\sqrt{p^3}}$

APPENDIX C
Mathematical Formulas: Geometry, Solid Bodies

Appendix C

TABLE C.1 Mathematical Formulas: Geometry, Solid Bodies

V = volume	A = cross-sectional area	A_s = surface area	A_m = generated surface
Cuboid	$V = a \cdot b \cdot c$ $A_s = 2(a \cdot b + a \cdot c + b \cdot c)$ $d = \sqrt{a^2 + b^2 + c^2}$	**Cone**	$V = \dfrac{\pi}{3} r^2 h$ $A_m = \pi r L$, $A_s = \pi r(r + L)$ $L = \sqrt{r^2 + h^2}$
Triangular Prism	$V = \dfrac{1}{3}(a + b + c) A$	**Frustum of Cone**	$V = \dfrac{\pi h}{3}(R^2 + r^2 + Rr)$ $A_m = 2\pi \cdot \rho \cdot L$ $\rho = 0.5(R + r)$ $L = \sqrt{(R^2 - r^2) + h^2}$
Pyramid	$V = \dfrac{A_1 h}{3}$	**Sphere**	$V = \dfrac{4}{3}\pi r^3 = 4.189 r^3$ $= \dfrac{1}{6}\pi d^3 = 0.5236 d^3$ $A_s = 4\pi r^2 = \pi d^2$
Frustum of Pyramid	$V = \dfrac{h}{3}(A_1 + A_2 + \sqrt{A_1 A_2})$	**Segment of a Sphere**	$V = \dfrac{\pi}{6} h \left(\dfrac{3}{4} s^2 + h^2 \right)$ $= \pi h^2 \left(r - \dfrac{h}{3} \right)$ $A_m = \dfrac{\pi}{4}(s^2 + 4h^2)$ $= 2\pi r h$
Cylinder	$V = \dfrac{\pi}{4} d^2 h$ $A_m = 2\pi r h$ $A_s = 2\pi r(r + h)$	**Sector of a Sphere**	$V = \dfrac{2}{3}\pi r^2 h$ $A_s = \dfrac{\pi}{2} r(4h + s)$

Mathematical Formulas: Geometry, Solid Bodies

TABLE C.1 Mathematical Formulas: Geometry, Solid Bodies (*Continued*)

V = volume	A = cross-sectional area	A_s = surface area	A_m = generated surface
Zone of a Sphere	$V = \dfrac{\pi}{6} h(3a^2 + 3b^2 + h^2)$ $A_s = \pi(2rh + a^2 + b^2)$ $A_m = 2\pi rh$	Ungula	$V = \dfrac{2}{3} r^2 h$ $A_s = A_m$ $+ \dfrac{\pi}{2}\left(r^2 + r\sqrt{r^2 + h^2}\right)$ $A_m = \pi dh$
Sliced Cylinder	$V = \dfrac{\pi}{4} d^2 h$ $A_s = \pi r \left[h_1 + h_2 + r \right.$ $\left. + \sqrt{r^2 + (h_1 - h_2)^2/4}\,\right]$ $A_m = \pi dh$	Barrel	$V = \dfrac{\pi}{12} h(2D^2 + d^2)$

Plane Analytic Geometry (Equations)

Straight Line	$y = mx + b$ $m = \dfrac{y_2 - y_1}{x_2 - x_1} = \tan\phi$	Circle	$(x-a)^2 + (y-b)^2 = r^2$ If $a=0$, $b=0$: $x^2 + y^2 = r^2$
Ellipse	$\dfrac{x^2}{a^2} + \dfrac{y^2}{b^2} = 1$ $c = \sqrt{a^2 - b^2}$ $\varepsilon = \dfrac{c}{a} < 1$	Hyperbola	$\dfrac{x^2}{a^2} - \dfrac{y^2}{b^2} = 1$ $c = \sqrt{a^2 + b^2}$ $\varepsilon = \dfrac{c}{a} > 1$
Parabola	$x^2 = 2py$ $OF = \dfrac{p}{2}$	Parabolic Arch	$y = \dfrac{4f}{L^2} x(L - x)$

APPENDIX D
Mathematical Formulas: Trigonometry

TABLE D.1 Mathematical Formulas: Trigonometry

Basic Conversions			
$\tan \alpha = \dfrac{\sin \alpha}{\cos \alpha}$	$\sec \alpha = \dfrac{1}{\cos \alpha}$	$\sin^2 \alpha + \cos^2 \alpha = 1$	$\dfrac{1}{\cos^2 \alpha} = 1 + \tan^2 \alpha$
$\cot \alpha = \dfrac{\cos \alpha}{\sin \alpha}$	$\operatorname{cosec} \alpha = \dfrac{1}{\sin \alpha}$	$\tan \alpha \cdot \cot \alpha = 1$	$\dfrac{1}{\sin^2 \alpha} = 1 + \cot^2 \alpha$
$\sin(\alpha \pm \beta) = \sin \alpha \cdot \cos \alpha \pm \cos \alpha \cdot \sin \beta$		$\tan(\alpha \pm \beta) = \dfrac{\tan \alpha \pm \tan \beta}{1 \mp \tan \alpha \cdot \tan \beta}$	
$\cos(\alpha \pm \beta) = \cos \alpha \cdot \cos \beta \mp \sin \alpha \cdot \sin \beta$		$\cot(\alpha \pm \beta) = \dfrac{\cot \alpha \cdot \cot \beta \mp 1}{\cot \beta \pm \cot \alpha}$	
$\sin 2\alpha = 2 \sin \alpha \cdot \cos \alpha$		$\tan 2\alpha = \dfrac{2 \tan \alpha}{1 - \tan^2 \alpha}$	
$\cos 2\alpha = \cos^2 \alpha - \sin^2 \alpha$		$\cot 2\alpha = \dfrac{\cot^2 \alpha - 1}{2 \cot \alpha}$	
$\sin 3\alpha = 3 \sin \alpha - 4 \sin^3 \alpha$		$\tan 3\alpha = \dfrac{3 \tan \alpha - \tan^3 \alpha}{1 - 3 \tan^2 \alpha}$	
$\cos 3\alpha = 4 \cos^3 \alpha - 3 \cos \alpha$		$\cot 3\alpha = \dfrac{\cot^3 \alpha - 3 \cot \alpha}{3 \cot^2 \alpha - 1}$	
$\sin \dfrac{\alpha}{2} = \sqrt{\dfrac{1 - \cos \alpha}{2}}$		$\tan \dfrac{\alpha}{2} = \dfrac{\sin \alpha}{1 + \cos \alpha} = \dfrac{1 - \cos \alpha}{\sin \alpha} = \sqrt{\dfrac{1 - \cos \alpha}{1 + \cos \alpha}}$	
$\cos \dfrac{\alpha}{2} = \sqrt{\dfrac{1 + \cos \alpha}{2}}$		$\cot \dfrac{\alpha}{2} = \dfrac{\sin \alpha}{1 - \cos \alpha} = \dfrac{1 + \cos \alpha}{\sin \alpha} = \sqrt{\dfrac{1 + \cos \alpha}{1 - \cos \alpha}}$	
$\sin \alpha = 2 \sin \dfrac{\alpha}{2} \cdot \cos \dfrac{\alpha}{2} = \dfrac{2 \tan \dfrac{\alpha}{2}}{1 + \tan^2 \dfrac{\alpha}{2}}$		$\tan \alpha = \dfrac{2 \tan \dfrac{\alpha}{2}}{1 - \tan^2 \dfrac{\alpha}{2}}$	
$\cos \alpha = \cos^2 \dfrac{\alpha}{2} - \sin^2 \dfrac{\alpha}{2} = \dfrac{1 - \tan^2 \dfrac{\alpha}{2}}{1 + \tan^2 \dfrac{\alpha}{2}}$		$\cot \alpha = \dfrac{\cot^2 \dfrac{\alpha}{2} - 1}{2 \cot \dfrac{\alpha}{2}}$	
$\sin \alpha + \sin \beta = 2 \sin \dfrac{\alpha + \beta}{2} \cdot \cos \dfrac{\alpha - \beta}{2}$		$\cos \alpha + \cos \beta = 2 \cos \dfrac{\alpha + \beta}{2} \cdot \cos \dfrac{\alpha - \beta}{2}$	
$\sin \alpha - \sin \beta = 2 \cos \dfrac{\alpha + \beta}{2} \cdot \sin \dfrac{\alpha - \beta}{2}$		$\cos \alpha - \cos \beta = -2 \sin \dfrac{\alpha + \beta}{2} \cdot \sin \dfrac{\alpha - \beta}{2}$	
$\tan \alpha \pm \tan \beta = \dfrac{\sin(\alpha \pm \beta)}{\cos \alpha \cdot \cos \beta}$		$\cot \alpha \pm \cot \beta = \dfrac{\sin(\beta \pm \alpha)}{\sin \alpha \cdot \sin \beta}$	

TABLE D.1 Mathematical Formulas: Trigonometry (Continued)

Basic Conversions	
$\sin\alpha \cdot \cos\beta = \frac{1}{2}\sin(\alpha+\beta) + \frac{1}{2}\sin(\alpha-\beta)$	$\tan\alpha \cdot \tan\beta = \frac{\tan\alpha + \tan\beta}{\cot\alpha + \cot\beta}$
$\cos\alpha \cdot \cos\beta = \frac{1}{2}\cos(\alpha+\beta) + \frac{1}{2}\cos(\alpha-\beta)$	$\cot\alpha \cdot \cot\beta = \frac{\cot\alpha + \cot\beta}{\tan\alpha + \tan\beta}$
$\sin\alpha \cdot \sin\beta = \frac{1}{2}\cos(\alpha-\beta) - \frac{1}{2}\cos(\alpha+\beta)$	$\cot\alpha \cdot \tan\beta = \frac{\cot\alpha + \tan\beta}{\tan\alpha + \cot\beta}$
$\sin^2\alpha - \sin^2\beta = \sin(\alpha+\beta)\cdot\sin(\alpha-\beta)$	$\cos\alpha + \sin\alpha = \sqrt{2}\cdot\sin(45°+\alpha)$
$\cos^2\alpha - \sin^2\beta = \cos(\alpha+\beta)\cdot\cos(\alpha-\beta)$	$\cos\alpha - \sin\alpha = \sqrt{2}\cdot\cos(45°+\alpha)$

$\alpha°$	0°	30°	45°	60°	90°
α (rad)	0.0	$\frac{\pi}{6} = 0.5236$	$\frac{\pi}{4} = 0.7854$	$\frac{\pi}{3} = 1.0472$	$\frac{\pi}{2} = 1.5708$
$\sin\alpha$	0.0	$\frac{1}{2} = 0.5000$	$\frac{\sqrt{2}}{2} = 0.7071$	$\frac{\sqrt{3}}{2} = 0.8660$	1.0
$\cos\alpha$	1.0	$\frac{\sqrt{3}}{2} = 0.8660$	$\frac{\sqrt{2}}{2} = 0.7071$	$\frac{1}{2} = 0.5000$	0.0
$\tan\alpha$	0.0	$\frac{\sqrt{3}}{3} = 0.5774$	1.0	$\sqrt{3} = 1.7321$	$\pm\infty$
$\cot\alpha$	$\pm\infty$	$\sqrt{3} = 1.7321$	1.0	$\frac{\sqrt{3}}{3} = 0.5774$	0.0

ϕ	$-\alpha$	$90° \pm \alpha$	$180° \pm \alpha$	$270° \pm \alpha$	$360° - \alpha$
$\sin\phi$	$-\sin\alpha$	$+\cos\alpha$	$\mp\sin\alpha$	$-\cos\alpha$	$-\sin\alpha$
$\cos\phi$	$+\cos\alpha$	$\mp\sin\alpha$	$-\cos\alpha$	$\pm\sin\alpha$	$+\cos\alpha$
$\tan\phi$	$-\tan\alpha$	$\mp\cot\alpha$	$\pm\tan\alpha$	$\pm\cot\alpha$	$-\tan\alpha$
$\cot\phi$	$-\cot\alpha$	$\mp\tan\alpha$	$\pm\cot\alpha$	$\pm\tan\alpha$	$-\cot\alpha$

APPENDIX E
Symbols

Table E.1 Symbols

Symbol	Description	Symbol	Description
A	Area, cross-sectional area (cm²)	R	Support reaction (kN), strength (MPa), radius (cm)
D	Diameter (cm), force (kN)		
E	Modulus of elasticity (MPa) For steel: $E = 2 \times 10^5$ MPa	S	Settlement (cm)
		S	Elastic section modulus about neutral axis (cm³)
E_s	Modulus of deformation of soil (MPa)		
F_c	Centrifugal force (kN)	S_x	Elastic section modulus about x-x axis (cm³)
G	Shear modulus of elasticity (MPa)	S_y	Elastic section modulus about y-y axis (cm³)
	For steel: $G = 77{,}221$ MPa	S_z	Elastic section modulus about z-z axis (cm³)
H	Horizontal support reaction (kN)	T	Temperature (°C, °F)
I	Moment of inertia of section about neutral axis (cm⁴)	V	Shear (kN), volume (cm³, m³)
		W	Weight (kN)
I_x	Moment of inertia of section about x-x axis (cm⁴)	Z	Plastic section modulus (cm³), force (kN)
I_y	Moment of inertia of section about y-y axis (cm⁴)	c	Cohesion (Pa)
I_z	Moment of inertia of section about z-z axis (cm⁴)	e	Eccentricity (cm)
I_p	Polar moment of inertia (cm⁴)	g	Gravitational acceleration ($g = 9.81$ m/s²)
K_0	Coefficient of earth pressure at rest	i	Radius of gyration (cm)
K_a	Coefficient of active earth pressure	k_w	Winkler's coefficient of subgrade (kN/cm³)
K_p	Coefficient of passive earth pressure	n	Porosity (%)
K_{aE}	Coefficient of seismic active earth pressure	p	Horizontal distributed load (kN/m)
L	Span length (m)	w	Vertical distributed load (kN/m)
M	Mass (kg)	σ	Direct stress (Pa)
M	Bending moment about neutral axis (kN·m)	τ	Shear stress (Pa)
		τ_s	Shear strength (Pa)
M_x	Bending moment about x-x axis (kN·m)	γ	Unit volume weight (kN/m³)
M_y	Bending moment about y-y axis (kN·m)	μ	Poisson's ratio
M_z	Bending moment about z-z axis (kN·m)	α	Coefficient of linear expansion (1/grad)
M_D	Dynamic bending moment (kN·m)	ρ	Unit mass (kg)
N	Axial force (kN)	Δ	Deflection (cm)
P	Applied load (kN)	ϕ	Angle of internal friction
P_e	Euler's force (kN)	$\tan\phi$	Coefficient of friction

Index

A

Allowable bearing pressures in soils, 158, 166
   bearing capacity analysis, 164–165
   calculation for, 158
   cohesionless soils, 160
   cohesive soils, 160
   elastic theory, 158
Allowable loads on piles, 170
Analysis, of frame, 84
Anchored sheet pile walls, 203
Angle of deflection of beams, 51
Angular deflections, 34
Arches, 93–115
   crescent beam position stress factors, 108
   eccentrically curved beams, 106
   fixed parabolic arches, 105, 107, 110–111
   formulas for circular rings and arches, 98, 102
   influence lines, 111
   length of cable carrying known loads, 114–115
   reactions of a three-hinged arch, 112
   steel rope, deflection, forces, temperature, 113

Arches (*Cont.*):
   support reactions and bending moments, 95–111
   symmetrical three-hinged arch, 96
   two-hinged parabolic arch, 100–101, 103

B

Beam loading formulas, 38
   angle of deflection, 51
   beams fixed at both ends, 65
Beams, diagrams and formulas for various loading conditions, 47–80
   bending moment, 51
   bending moments and deflection, 63
   cantilever beams, 57
   characteristics of loadings, 64
   coefficients for correcting values, 54
   combined axial and bending loads, 60
   computation of fixed-end moments in prismatic beams, 62–64
   computation of simple beam, 50
   continuous beams, 66–67
   curved beams, 72, 74

Beams, diagrams and formulas for various loading conditions (*Cont.*):
   deflection, 51, 52–53
   eccentric loading, 58
   fixed at one end, 59
   fixed at one end, supported at other end, 61
     greatest safe load, 52–53
   influence lines, 73, 75–79
     load distributed, 52–53
     load in middle, 52–53
     loadings, 51, 63
     moving concentrated loads, 70–71
   natural circular frequencies of vibration of prismatic beams, 80
   overhanging one support, 55
   safe loads for beams, 52–53
     section shape, 52–53
   settlement of beam support, 68–69
   torsion in structural members, 56
Bearing plates, 136
Bending moment of beams, 51, 63
Bending, strain energy in, 12

C

Cable carrying a known load, 114–115
Cantilever beams, 57, 64
   retaining walls, 192, 199
   sheet piling, 201
     calculation of, 202
Circular cross section pipes and tunnels, 217, 219, 221
Circular plates, 145
Circular rings and arches, 98, 102
Column base plates, 132
Columns, 86, 92
   buckling formulas for, 92
   elastic flexural buckling of, 90
   short, 88

Combined axial and bending loads in beams, 60
Computation of truss stress, 124
   in truss diagonal, 122
Concrete gravity retaining walls, 196, 198
Conduit, ditch, loads on, 212, 214, 216
Continuous beams, 66–67
Conversion tables, 226–234
   Angle-American and metric system, 233–234
   quick-use, 226–232
Crescent beam, 108
Curved beams, 51–53, 72, 74, 106
Curved springs, 22

D

Direct foundations, 169–170
Ditch piping, loads on, 212, 214, 216

E

Earthquake equation, 187
Eccentric loading of beams, 58
Elastic design, 25
Energy, strain, 8–12
Engineering properties of soils, 153

F

Failure analysis, 18
Fixed parabolic arches, 105, 107, 110–111
Fixed-end prismatic beams, 62–64
Flange plate thickness, 140
Flat metal springs, 28
Flat plates, 144–147
Flow of water in soils, 155
Forced oscillation of beams, 23–24
Formulas, mathematical, 236–241
   algebraic, 236–237
   geometry, 240–241
   solid bodies, 240–241
   trigonometry, 244–245

Index **251**

Foundations, 167–182
   allowable loads on piles, 170
      Boussinesq equation, 176
      calculation of, 170
      contact pressure on, 169
   direct foundations, 169–182
      distribution of loads, 171, 173
      estimate of settlement, 178
      footing size, 182
      group capacity, 173
      Housel's method for, 182
      one-way action of, 169
      pile capacity, 175
      rigid continuous beam, 177
      rigid continuous footing, 180–181
      settlement, determination of, 174, 176
      stability of, 171
      substructures and superstructures, 172
      toe capacity load, 172
Frames for static loading conditions, 81–92
   analysis of frame, 84
   buckling formulas for columns, 92
   columns and frames, 86
   diagrams for static loading, 83, 85, 87, 89, 91
   elastic flexural buckling of columns, 90
   short columns, 88
Geometric sections, 29–43
   angular deflection, 34
   beam loading formulas, 38
   column characteristics, 30
   position of flexural center, 40
   shaft twist and torque formulas, 36
   torsion in solid and hollow shafts, 42
   torsion of shafts, 34
   various cross sections, 31–33

H

Housel's method, 182

I

Influence lines, 73, 75–79, 123, 125
   load distributed, 52–53
   load in middle, 52–53
   loadings, 51, 63

L

Lateral pressures in soils, 156
Local buckling of plates, 134

M

Mathematical formulas, 236–245
   algebra, 236–237
   geometry, 240–241
   solid bodies, 240–241
   trigonometry, 244–245
Method of joints, 117–125
   of section analysis, 119
Metric system, conversion tables, 226–234
Modulus of deformation of soils, 152, 163
Moving concentrated loads, 70–71
Moving loads, stresses produced by, 128

N

Natural circular frequencies of vibration of prismatic beams, 80
   overhanging at one support, 55

P

Piles, 170–175
   capacity of, 175
   distribution of loads, 171, 173
   group capacity, 173
   toe capacity, 172
Pipe bends, forces due to, 220

Pipe stresses, perpendicular to longitudinal axis, 218
Pipes and tunnels, 205–223
   calculation of rectangular pipe 210
   circular cross section, 217, 219, 221
   forces due to pipe bends, 220
   loads on ditch piping and conduit, 212, 214
   pipe on supports at intervals, 222
   pipe stress perpendicular to the longitudinal axis, 218
   rectangular cross section, 207, 209, 211, 213, 215
Plates, 127–147
   bearing plates, 136
   circular, 145
   column base plates, 132
   computation of rectangular plates, 130, 138
   flange plate thickness, 140
   flat, 144–147
   formulas for flat plates, 144–147
   local buckling of plates, 134
   rectangular, 127–143
   stresses in plates, 142
Poisson's ratio, 152

Q

Quick-use conversion formulas, 226–234

R

Rectangular cross section pipes and tunnels, 207, 209, 211, 213, 215
Rectangular plates, 127–143
Retaining structures, 185–204
   anchored sheet pile walls, 203
   calculation of, 188, 190
   cantilever retaining walls, 192, 199
   calculation of, 200

Retaining structures (*Cont.*):
   cantilever sheet piling, 201
   calculation of, 202
   concrete gravity retaining walls, 196, 198
   earthquake equation for, 187
   geosynthetics in wall construction, 194
   lateral earth pressure on retaining walls, 187, 191, 193, 195, 197
   earth pressure on, 191
   examples of, 189
   stability of a retaining wall, 204

S

Safe loads for beams, 52–53
   section shape, 52–53
Settlement of beam support, 68–69
Settlement of soils, 158
Shafts, torsion in solid and hollow, 34, 42
   twist and torque formulas, 36
   various cross sections, 31–33
Shear, strain energy in, 8
Shear strength of soils, 163
Simple beams, 49–53
Slope stability analysis, 163
Soils, 151–166
   allowable bearing pressures, 166
   alternative formulas, 158
   bearing capacity analysis, 164–165
   calculation for, 158
   cohesionless soils, 160
   cohesive soils, 160
   elastic theory, 158
   engineering properties, 153
   flow of water in soil, 155
   forces on retaining walls, 156
   lateral pressures in, 156
   modulus of deformation, 152, 163
   Poisson's ratio, 152

Soils (*Cont.*):
   settlement of, 158
   shear strength of, 163
   slope stability analysis, 163
   vertical pressures in, 156
      approximate method, 156
      concentrated load, 156
      uniformly distributed load, 156
   weights and volumes, 154–155
   Winkler's hypothesis, 161–163
Stability of a retaining wall, 204
Static loading, frames for, 81–92
Steel rope, deflection, forces, temperature, 111
Strain energy, 8, 10, 12
   in bending, 12
   in shear, 8
   in structural members, 10, 12
   in torsion, 12
Stress and strain, 4–28
   bending, 7, 9, 11
   buffer spring and column, 27
   continuous deep beams, 17
   curved springs, 22
   elastic design, 25
   failure analysis, 18
   flat metal springs, 28
   forced oscillation of beams, 23–24
   methods of analysis, 27
Stresses in plates, 142
Structures, retaining, 185–204
Support reactions and bending moments of arches, 95–111
Symbols, for this book, 248

T

Tables, conversion, 226–234
Tension and compression, 5
   torsion, 13
   transverse bending, 15
   transverse oscillation in beams, 19, 21
Three-hinged arch, 112
Torsion, strain energy in, 12
   in structural members, 56
Transverse bending, 15
   oscillation in beams, 19, 21
Trusses, 117–126
   computation of truss, 124
   determining stress in truss diagonal, 122
   influence lines, 123, 125
   method of joints, 117–125
   method of section analysis, 119
   stresses produced by moving loads, 128
   types of trusses, 120
Two-hinged parabolic arch, 100–101, 103

V

Vertical pressures in soils, 156

W

Weights and volumes of soils, 154–155
Winkler's hypothesis, 161–163